Science, Technology,
and Politics

Science, Technology, and Politics
Policy Analysis in Congress

EDITED BY
Gary C. Bryner

Routledge
Taylor & Francis Group

LONDON AND NEW YORK

First published 1992 by Westview Press

Published 2019 by Routledge
52 Vanderbilt Avenue, New York, NY 10017
2 Park Square, Milton Park, Abingdon, Oxon OX14 4RN

Routledge is an imprint of the Taylor & Francis Group, an informa business

Copyright © 1992 by Taylor & Francis

Library of Congress Cataloging-in-Publication Data
Science, technology, and politics : policy analysis in Congress /
    edited by Gary C. Bryner.
        p. cm.
    Includes bibliographical references and index.
    1. Science and state—United States. 2. Technology and state—
United States. 3. United States. Congress. I. Bryner, Gary C.,
1951-
Q127.U6S3275 1992
338.97306—dc20   92-371
                                                                        CIP

ISBN 13: 978-0-367-28676-7 (hbk)
ISBN 13: 978-0-367-30222-1 (pbk)

To Benjamin, future scientist

Contents

Acknowledgments

This book began several years ago as a project organized by members of the Science and Technology Studies section of the American Political Science Association. It is part of an ongoing attempt by members of the section and others to focus scholarly attention on the political and social implications of technological change and scientific advances. Part of the concern is to identify theories, conceptual frameworks, and concepts from political science that can usefully be applied to the study of science and technology. Part of the concern is to explore how science and technology-related concerns help illuminate and test some of the enduring theories of political science. We hope to contribute to the development of a strong theoretical underpinning for science and technology studies. We hope that such an enrichment of the theoretical bases for understanding science and technology-related phenomena will also contribute to more effective and appropriate public policies for regulating and encouraging scientific and technological developments. This book is an attempt to marry theoretical exposition and applied policy inquiry.

We are greatly indebted to a number of people who provided interviews, read drafts, and participated in meetings. Individual authors have indicated particular sources of support. Financial support for producing the volume was provided by Brigham Young University. We are particularly grateful to officials from the U.S. Office of Technology Assessment for their willingness to meet with us and discuss their views of policy analysis in a 1989 workshop at Georgia Tech University. We also greatly appreciate the opportunity to work with Amy Eisenberg and others at Westview Press. Their professionalism and interest in our work has made the entire project a most interesting and rewarding enterprise.

Gary C. Bryner

About the Editor
and Contributors

Robert H. Blank is professor of political science, University of Canterbury, Christchurch, New Zealand. He is the author of several books and articles, including *Life, Death, and Public Policy*.

Gary C. Bryner is associate professor of political science, Brigham Young University, and author of *Bureaucratic Discretion, In Search of the Republic*, and a forthcoming book on the new Clean Air Act.

W. D. Kay is an assistant professor of political science at Northeastern University. He has written a number of articles on large-scale technologies and is currently writing a book on the politics of technological development.

William W. Keller is a senior analyst and project director at the Office of Technology Assessment, United States Congress. His research focuses on multinational corporations, global arms trade, and defense technology. He is the author of *The Liberals and J. Edgar Hoover: Rise and Fall of a Domestic Intelligence State* and of a forthcoming book on state sovereignty and the proliferation of modern weapons.

Michael E. Kraft is professor of political science and public affairs and Herbert Fisk Johnson Professor of Environmental Studies at the University of Wisconsin, Green Bay. He is coeditor and contributing author of a number of books, including *Environmental Policy in the 1990s, Technology and Politics*, and *Public Opinion and Nuclear Waste*.

L. Christopher Plein is a Ph.D. candidate in political science at the University of Missouri–Columbia and is conducting research on the evolution of biotechnology as a policy issue. His larger research concerns include social responses to new technology and global economic developments.

Priscilla M. Regan is assistant professor in the Department of Public Affairs, George Mason University. She formerly worked as a senior analyst at the U.S. Congress Office of Technology Assessment and is currently writing a book on privacy, technology, and public policy.

Norman J. Vig is professor of political science at Carleton College. He is coeditor and contributing author of a number of books, including

Environmental Policy in the 1990s and *Technology and Politics*. His current research focuses on environmental policy in the European Community.

Kathryn D. Wagner is a program officer in the Conservation and Environment Program, Pew Charitable Trusts. She was an analyst with the U.S. Congress Office of Technology Assessment and was an assistant professor of political science at Swarthmore College.

David J. Webber is associate professor of political science at the University of Missouri–Columbia and specializes in American public policy. His current research interests are policy makers' use of policy information and biotechnology policy. He recently edited *Biotechnology: Assessing Social Impacts and Policy Implications*.

E. J. Woodhouse is a professor in the Science and Technology Studies Department, Rensselaer Polytechnic Institute. His research focuses on policy-analytic strategies for promoting more intelligent and more democratic policy making. His books include *The Policy-Making Process*, third edition, with Charles E. Lindblom and *The Demise of Nuclear Energy: Lessons for Democratic Control of Technology* with Joseph Morone.

PART ONE

Introduction

1

Science, Technology, and Policy Analysis in Congress: An Introduction

Gary C. Bryner

Introduction

Developments in science and technology pose profound challenges for humankind. The range of issues and the velocity at which change occurs are overwhelming. Air pollution, produced as a by-product of dramatic technological innovations in industry and transportation, threatens the health of millions of human beings. Projected changes in the global climate could disrupt agricultural activity, endanger water supplies, cause widespread flooding, and jeopardize the survivability of natural habitats and biodiversity. Medical and other wastes contaminate the aquatic food supply and imperil human health as they wash up on beaches.

New reproductive technologies are redefining what it means to be a human being. The future genetic pool of the human race may be altered as a result of efforts such as the mapping of the human genome. Genetic research promises eventually to eliminate diseases that have ravaged millions of lives over the centuries. Biomedical advances have made it possible to maintain lives that would otherwise be lost, but the enormous costs of these technologies and the way in which they are used pose perplexing questions. Developments in biotechnology raise fundamental concerns about the safety of our food supply and the consequences of biological processes becoming commercialized.

Traditional values of privacy and individual autonomy are threatened by the development of new communication technologies. The ability of government and private institutions to collect and process information on individuals has exploded during the past three decades. Rights of privacy, due process, and protection against unreasonable searches that have long been central to American constitutional democracy are threatened by the growth of national computerized data bases.

The future of the American economy and the quality of life enjoyed by Americans has become increasingly intertwined with technological developments. The competitiveness of U.S. industries in global markets is to a great extent a function of their ability to develop and market new technologies. Major scientific research projects cost billions of dollars and raise fundamental questions about the ability of the federal government to finance such efforts in an era of budget deficits and declining discretion in new spending. Space exploration, the search for new forms of energy production, and national defense weapons technologies generate concerns about how much of these innovations we can afford to pursue and what areas should be left to private funding.

Public policies that promote or regulate scientific research and technological innovations are among the most important activities undertaken by governments. These policies share a number of characteristics that make policy making for them particularly challenging. In some cases, the consequences of policy action or inaction have enormous consequences and are largely irreversible. If the projected increase in global temperatures are realized, for example, the consequences will likely be catastrophic and beyond the control of most of the inhabitants of the earth.

The distribution of the consequences of technological advances pose great challenges to democratic policy making. Many of the adverse consequences of activities such as the production of nuclear energy will fall on future generations who will have to deal with radioactive wastes while the benefits are largely confined to the current generation. The poorest countries are least able to protect their populations against the problems of pollution and environmental degradation. Efforts to remedy the environmental consequences of technological changes require cooperation and coordination. It is not clear how the interests of future generations or subgroups of the population that have the fewest economic and political resources can be protected in a political system dominated by well-heeled interests.

Science and technology policies are particularly complex. Policy choices are permeated by uncertainty. Decisions must be made by political actors, but they must be based on the latest scientific research. However, there is rarely consensus in the scientific community and policy makers must function within a context of uncertainty and disagreement among experts. The nature of scientific debate and research, the slowness in which the scientific process often proceeds, and the tentativeness of scientific findings clash with short-run orientation of politics and the demand for unambiguous solutions to pressing problems. These issues cross traditional policy boundaries and disciplines. Scientific and technological expertise compete with that surrounding established substantive policy specializations.

Central to the policy making process is the ability of policy makers to assess the strengths and weaknesses of alternative policy actions and to choose those that are most consonant with policy makers' values and priorities. Policy analysis in general rests on the expectation that the technical assessment of competing policy options will be separated from the political calculations of the policy makers—that there will be an objective, non-political assessment of policy options before the inevitable political calculations shape the decisions eventually made. Careful policy analysis will precede the application of narrow political pressure and ensure that policies producing the greatest net gains in social welfare will be pursued (Stokey and Zeckhauser 1978).

Policy analysis is not without its critics, however. The reliance of most analytic techniques on measures of economic efficiency and utility may clash with other values such as distributive justice (Rawls 1971; Nozick 1974). Reliance on analytic techniques may give the illusion of precision, certainty, and objectivity, when in fact decisions must be made on much more subjective grounds. Policy analysis may enhance the role of experts at the expense of elected officials, thus reducing the accountability of policy making to democratic forces (Hawkesworth 1988; Sartori 1962).

Some have entered the debate over the advantages and disadvantages of policy analysis by arguing that it is simply irrelevant, that analyses have little impact on policy making. Analysis is often viewed as little more than a political tool used for political purposes such as lending legitimacy to policy choices based on other criteria. Policy analysis is limited to a symbolic rather than a substantive role in policy making (Straussman 1978).

The purpose of this book is to examine how the assessment of science and technology-related policies takes place in Congress, the role it plays in the formulation of public policies, and the prospects for improving policy making for these kinds of policies. The assessment of science and technology policies is similar to policy analysis in general as it seeks to illuminate the causes of a public problem and the associated values and issues, identify and compare options for remedying the problem, and recommend what policy steps should be pursued. We are convinced that the challenges posed by science and technology-related policies are particularly difficult and deserve special attention, and that conclusions that are drawn here can also shed light on issues surrounding policy analysis in general. These policies are of primary importance and policy makers need to be able to be able to assess effectively alternative solutions while recognizing that policy making for these kinds of policies can threaten democratic values of accountability and representation.

Congress and the Making of Science
and Technology Policy

The way in which Congress assesses scientific and technological issues deserves particular attention for several reasons. These issues combine scientific uncertainty over causes and consequences, major economic factors, and primary political and cultural values into a complex web of policy choices that reach the most fundamental concerns in American society. The executive branch of government, many believe, is best suited to make these kinds of choices. A permanent bureaucracy, the reliance on specialists and expertise, the size and range of jurisdictions and responsibilities, an orientation toward long-run considerations, and a chief executive who serves as a focal point for policy coordination are all part of the idea that making science and technology policy should largely be the purview of the executive branch.

Congress, however, is at the center of governmental efforts to promote and regulate economic, industrial, and commercial activity. Congress plays a fundamental role in authorizing by statute significant promotional and regulatory actions of agencies and departments, and congressional appropriations help determine priorities and allocate resources. A legislative body, charged with responsibility to represent competing geographic regions, is faced with particular difficulties in assessing policies concerning scientific and technological developments. The bargaining and compromising that are central to congressional politics sometimes clash with the scientific world of careful testing of hypotheses and the patient development of consensus. Electoral incentives demand that members of Congress focus much of their attention on short-run, immediate benefits and burdens affecting constituents, while science prefers tentativeness and caution in the face of uncertainty and the development of long-term strategies to remedying problems. The time required to assess thoroughly the consequences of technological change are often much greater than the pressures of politics will permit. The decentralized structure of Congress produces a great variety of policy responses that pose challenges for priority-setting and comprehensive policy making.

How well does Congress deal with the challenge of assessing policy options involving complex scientific processes and technologies? Some critics have argued that members of Congress do not have sufficient knowledge to make decisions concerning the regulation of industrial and technological processes and their outcomes or that it is too dependent on the executive branch for this kind of analysis. Others warn that congressional institutions and processes are not well suited to make policies based on these kinds of issues and new institutions are needed. Much of the debate focuses on whether Congress should increase efforts to govern technology or whether private and market-oriented institutions and mechanisms are better suited for such tasks. Despite a recognition of the tremendous uncertainty associated with technological developments, commentators often argue that Congress has done a poor job of anticipating future challenges to government and society resulting from scientific and technological advances.

In response to these and other criticisms, Congress has undertaken a number of efforts to improve its policy making for science and technology-related issues. Committees in both chambers have been created with jurisdiction over these matters. Congress has funded a tremendous increase in staff members, including scientists, engineers, and others with technical expertise. One of its most important acts was creating the Office of Technology Assessment in 1972, bolstering considerably its analytic capabilities. Congress has also strengthened the General Accounting Office and the Congressional Research Service, and established a number of advisory commissions and special committees to give advice and formulate policy options. Despite these impressive commitments, Congress confronts formidable challenges in choosing among policy options.

Policy Analysis and the Policy Making Process

Policy analysis plays a central role in all phases of the policy making process, particularly at the critical stage when Congress chooses among competing policy options and formulates the policies to be pursued (Brewer and deLeon 1986). The analysis of scientific and technological developments raises especially difficult questions, as indicated below. The chapters that follow explore these questions as they examine how well policy analysis has taken place at each step in the policy process and for a number of different policy areas.

Policy Initiation

The first step in the process, policy initiation involves the recognition of the existence of a problem and efforts to define its causes and consequences. The factors that determine which issues are placed on the congressional policy making agenda are critical. How problems are defined and what causes are attributed to them play a major role in determining what kinds of responses are pursued. If the causes of problems are not well understood, ameliorative policies may be misguided or inappropriate. What are the fundamental characteristics of science and technology-related policies? What particular challenges do they pose for policy making?

Policy Estimation

Involving the initial effort to outline possible policy responses and to make a preliminary assessment of the risks, costs, and benefits of alternative responses, this screening process will necessarily eliminate some alternatives. The analysis will also likely identify options that are the most promising candidates for action. Political forces will coalesce around these initial policy options, creating pressures and incentives for particular actions to be pursued. Why do some technology-related issues get on the political agenda and others do not? What policy options did Congress consider? Which interests were well represented? Not well represented? What issues were not adequately addressed?

Policy Selection

The selection process requires members of Congress to choose among competing policy options. These choices are a product of the interaction of the analysis of policy options and the political forces surrounding options. Such choices are largely political ones, but are assumed to be infused with a careful analysis of the strengths and weaknesses of competing alternatives. Why did Congress take the actions it did? Where did Congress get its information to generate policy options? How well did the congressional support agencies and other institutions and actors contribute to Congress' ability to assess the consequences of technological changes and develop appropriate policy responses? What were the values, assumptions, and biases underlying the assessments? How well did Congress respond to changes in technologies? How well were long-term, emerging problems assessed?

Policy Implementation

Primarily an administrative function, policy implementation also requires assessments of the risks, costs, and benefits of alternative means of achieving statutory mandates, unless they are explicitly spelled out by Congress. Even with specific, detailed laws agencies may be given competing policy goals by Congress or the president, or have insufficient resources to accomplish all that is required of them and must assess choices concerning what to do and how to do it. What happened as policies were implemented? Did Congress put in place an appropriate structure and provide adequate funding to achieve its goals? How well has Congress overseen the implementation and evaluated the effectiveness and appropriateness of the statutes it enacted?

Policy Evaluation

Often viewed as primarily an executive function, evaluation requires that the actual efforts and accomplishments of implementing agencies be compared with the policy goals established for them by Congress and the president and with the nature of the problems they are expected to solve. Congress also engages in such assessments as it oversees agency activities and prepares to reauthorize or amend enabling statutes and appropriate funds. Analysis focuses on how well the policy goals have been achieved as well as how well those goals reflect the nature of the problems they are designed to resolve. To what extent has Congress tried to assess the effectiveness of the policy choices it made? What role has policy analysis played as Congress has appropriated funds and reauthorized statutes? How can Congress' capacity to assess policy options be improved in ways that are consistent with other expectations of democratic government? How does policy making for policies affecting technological developments shed light on basic theories of democratic politics and government?

Kinds of Science and Technology Policies

Different kinds of policies are associated with different kinds of political arrangements. Science and technology-related policy issues are generally designed either to promote scientific and technological advances or to regulate these developments and their application. The federal government funds much of the research that takes place in the U.S. Congress is intimately involved in decisions concerning total spending levels and in the distribution of resources. To some extent, government sponsorship of research raises issues traditionally associated with distributive politics: will funds be spread widely throughout the nation, or concentrated in order to take advantages of economies of scale? Will Congress mandate where funds are to be spent, or will administering agencies be required to rely on competitive procedures and scientific peer review in making such decisions? Should large-scale projects be preferred over a larger number of smaller efforts? Should funding be limited to basic research, or include the development of new technologies that have marketable applications? Should the government ensure maximum dissemination of the results of federally funded research, or should it strengthen incentives for research by permitting researchers to receive financial gain from the commercialization of federally funded projects? How can Congress assess the merits of projects whose benefits are uncertain and, at best, may not be realized for several years?

A second category of policies are based on congressional efforts to regulate the use or consequences of technologies. Congress may place limits on what kinds of research can be done in areas such as genetic engineering. It may constrain the application of technologies that threaten constitutional protections of personal privacy or whose consequences are uncertain but potentially threatening. How can Congress assess the likely consequences of technological developments when these technologies are highly uncertain? How can Congress balance the possible benefits of permitting such developments with the possible harms resulting from them?

The Rest of the Book

The following chapters examine and compare different kinds of policies and the kinds of analytic and political factors associated with them. Chapter 2, by E. J. Woodhouse, provides an overview of the challenges posed to policy making by science-and-technology-related issues. It assesses two competing models of how policy analysis for these kinds of issues takes place, compares their strengths and weaknesses, and outlines the basic elements of an approach to policy analysis that is centered around a process for low-cost learning and flexible coping with uncertainty.

In Chapter 3, William W. Keller reviews the history and evolution of congressional responses to the challenges posed by developments in science and technology. This chapter provides an optimistic view on the potential for policy analysis to take place in Congress, but discusses some institutional barriers to effective policy making.

The chapters that follow in Part Two turn to how Congress and the federal government in general have formulated specific policies. In Chapter 4, Kathryn D. Wagner focuses on two policy areas—the Clean Air Act and the tracking and disposal of hazardous medical waste—and gives particular attention to how Congress has used scientific and technical information to address issues such as the assessment of risk, the balancing of costs and benefits in regulatory decisions, and scientific uncertainty in regulating air pollutants and medical wastes. The chapter emphasizes the differences between the Clean Air Act of 1970 and the Medical Waste Tracking Act enacted in 1988 that was shaped by twenty years of experience with environmental laws.

Chapter 5, by Michael E. Kraft, examines congressional efforts to deal with nuclear waste, population change and its consequences, and global climate change. Kraft gives particular attention to the importance of the role of policy entrepreneurs in policy making for new technologies and emphasizes that the most pressing challenge confronting Congress is not inadequate information but the need to improve its ability to make use of the policy analysis available to it.

Chapter 6, by W. D. Kay, traces congressional decisions to fund research for fusion energy and focuses on the issues of government funding of major scientific projects, the challenges associated with managing such projects, and the competing economic, political, and social values involved.

Chapter 7, by Robert H. Blank, centers on the current status of biomedical technology assessment and the challenges involved in assessing long-term problems, accounting for the interests of future generations, and dealing with the political constraints on policy assessment. Blank gives particular attention to the societal bias in favor of technological fixes to problems and the general unwillingness to make choices concerning the allocation of scarce resources.

Chapter 8, by L. Christopher Plein and David J. Webber, focuses on the production of knowledge concerning developments in biotechnology, the perceptions of Congress as they try to respond to this knowledge, and what kinds of knowledge need to be provided in order to produce better policy decisions. The chapter pursues these themes by examining several policy disputes: the use of bovine growth hormones, patents, commercialization of biotechnology, and the human genome mapping project.

Chapter 9, by Priscilla M. Regan, reviews how Congress has responded legislatively to developments in communication technologies and the competing values that policy makers have tried to satisfy, particularly the interaction of personal privacy and demands for efficient information collection and processing.

Chapter 10, by Gary C. Bryner, focuses on congressional efforts to strengthen the competitiveness position of U.S. industry in international markets. It examines the debate in Congress and the executive branch concerning the funding of basic versus applied research, large-scale versus smaller projects, and generic versus industry-specific assistance.

The chapters in Part Three compares technology policy assessment in the U.S. with other nations, draws some conclusions from the case studies presented here, and proposes areas for future research.

Chapter 11, by Norman J. Vig, places the experience of technology policy analysis in the U.S. in a comparative context by examining technology assessment in selected European nations. Such a comparative approach highlights the political differences between parliamentary systems and the American separation of powers and the implications for technology assessment.

The final chapter compares and integrates the analyses provided in the earlier chapters, explores ways that Congress might improve policy analysis for the scientific and technological challenges confronting it, and outlines some ideas for further research.

Bibliography

Brewer, Garry and Peter deLeon. 1983. *The Foundations of Policy Analysis*. Homewood, IL: The Dorsey Press.

Hawkesworth, M. E. 1988. *Theoretical Issues in Policy Analysis*. Albany, NY: SUNY Press.

Nozick, Robert. 1974. *Anarchy, State, and Utopia*. New York: Free Press.

Rawls, John. 1971. *A Theory of Justice*. Cambridge, MA.: Harvard University Press.

Sartori, Giovanni. 1962. *Democratic Theory*. Detroit, MI.: Wayne State University Press.

Stokey, Edith and Richard Zeckhauser. 1978. *A Primer of Policy Analysis*. New York: W. W. Norton.

Straussman, Jeffrey. 1978. *The Limits of Technocratic Politics*. New Brunswick, NJ: Transaction Books.

2

Toward More Usable
Technology Policy Analyses

E. J. Woodhouse

Introduction

Now that the Office of Technology Assessment is widely considered a success in Washington, and emulative efforts are springing up in other nations, it is appropriate to think about the next steps in the evolving craft of technology assessment. OTA and other analytic agencies have learned how to produce highly credible reports to actually help elected officials and other political actors do significantly better at steering technology. However, it may be necessary to attune assessment efforts more closely with our rudimentary understanding of the ways that knowledge comes to be employed in political life.

This chapter analyzes some of the main tasks that must be confronted by those doing—and using—technology policy analysis. The most significant of these is the high level of factual uncertainty that usually characterizes development and diffusion of new technologies and plagues efforts to assess their social and environmental problems. In addition to trying to reduce uncertainties through better data gathering and information processing, is there a way that analysis can aim explicitly and directly at helping political participants cope with the problem of uncertainty?

A second, related problem is that time and attention always are scarce, both for analysts and for their audiences; so which topics should be studied, and which aspects of those topics? How can we judge what *not* to study? And when we do take on a project, what sort of help ought we be trying to provide?

Finally, conflicts among values, and among different people's values, suffuse political life. There is no felicific calculus capable of showing conclusively how tradeoffs should be made even for one individual, much less for a society; only partisan judgments can do so. How should analysts adjust their work to take this into account—to provide as much assistance as possible to partisans in making their individual judgments and in evolving a collective policy?

To determine how well contemporary technology policy analyses are performing these tasks, we would need to examine an immense collection of studies. Relevant work comes from the Congressional Research Service, National Academy of Sciences/National Research Council, Office of Technology Assessment, General Accounting Office, and many of the mission agencies. Analyses bearing on technology policy are conducted as well by nongovernmental researchers in interest groups, universities, and think tanks. Any such comprehensive review would require cooperation among a large network of scholars, and I do not attempt it here. Instead, I draw very selectively on a handful of studies, principally from OTA, to illustrate and help develop a set of guidelines for improved technology policy analysis. I believe these arbitrarily selected studies are representative of the larger universe, but other observers will need to check my interpretations against parts of the technology analysis domain with which they are familiar. Together we may be able to develop a better understanding of the aims that ought to guide our work, and to evolve a collective craft that comes somewhat closer to achieving those aims.

Uncertainty

The starting point for any realistic understanding of politics is to face up to the inevitability of uncertainty. Analysts sometimes seem to behave as if this fundamental difficulty of political life could be circumvented by sufficient information or logic. On reflection, we all recognize that uncertainty (and therefore disagreement) seldom can be avoided in any realm of political life. In decision making about technology, one of the main sources of disagreement is that risks typically cannot be calculated very exactly. In the case of lead, the Environmental Protection Agency over a period of years offered more than a dozen technical claims regarding the need to take lead out of gasoline. Every one of the claims was plausibly rebutted by Ethyl Corporation, the major manufacturer of lead additives for gasoline. Thus, EPA claimed that "Since lead has not been shown to have any biologically useful function in the body, any increase in body burden of lead is accompanied by an increased risk of human health impairment." Ethyl Corporation countered: "Experiments on animals seem to show that lead is, after all, an essential element. . . . (Moreover) years of experience with occupationally exposed groups show blood lead levels well in excess of those found in the normally exposed population to be perfectly safe" (Collingridge and Reeve 1986, 42–44). The result was that there was ample room for each side to build up a case, changing it as time goes on, as new reports supplement the stock of literature and in response to one's opponents' criticisms. With such opportunities, there is no possibility of one side losing to the other, having to accept on scientific grounds the technical case made out by its rival (Collingridge and Reeve 1986, 45).

Much the same can be said of nuclear power plant safety and many other health/safety/environment issues. The range of legitimate disagreement sometimes can be narrowed, but often not far enough to be of much use to policy makers. A majority of atmospheric scientists, for example, believe that there will be climate warming from combustion of fossil fuels and release of other greenhouse gases, but they disagree markedly regarding how much warming, how soon, and what should be done about it (Morone and Woodhouse 1986; Falk and Brownlow 1989; Lowe 1989). Even for debates we think of as settled, such as smoking and lung cancer, inspection of the scientific literature often reveals a lively minority viewpoint (Collingridge and Reeve 1986). Another aspect of uncertainty about risks concerns the uses to which a technology will be put, the magnitude and frequency of its use, and the side effects and linkages. Contemporary work on the history of technology amply documents how rarely these matters are accurately predicted (Drucker 1973; Stross 1989):

- The original DDT researchers did not think of using the chemical as a pesticide for controlling insects on crops;
- No one foresaw the massive social changes produced by television;
- In 1972, the Atomic Energy Commission predicted that the U.S. would have 1200 nuclear power plants by the year 2000—we will actually have about 110;
- Predictions about computing, in contrast, have grossly underestimated improvements in power and declines in price, and have overestimated the social consequences.

Technology policy somehow must adapt to these realities; new scientific and engineering developments, entrepreneurial activities by business, and consumer responses can be expected routinely to evolve unexpected combinations and extensions of whatever exists at the time a forecast is attempted (Ascher 1978). So egregious errors are inevitable, and some form of trial, error, and error correction will be the normal process of technology policy making.

While every thoughtful observer of the human condition understands quite a lot about factual uncertainties and value conflicts, standard models of policy analysis offer little in the way of systematic strategies for coping with these endemic political phenomena. As a result, technology policy analysis starts off partially crippled, since uncertainty and value conflicts are precisely what policy makers have to deal with. Can we do better?

Coping with Uncertainty

One crucial step is simply to orient analysis toward the task of coping. It is not too much to say that the main task of technology policy analysis is to explain to decision makers the tactics available for coping with uncertainty in a particular problem area. The task has four facets. First, figure out what the risks are that need to be protected against.[1] In the case of biotechnology, this would include the possibility that the business sector might fail to launch new startup companies sufficiently diverse to carve out niches in each facet of the evolving industry, thereby ceding excessive shares to other nations.

A second step is to envision a broad range of strategies and tactics that potentially could help avert each major risk. We rarely can estimate very precisely the likelihood of any particular problem arising, nor can its magnitude be foretold. Early on there was no reliable way of predicting whether any given biotechnology product, company, or type of product would actually become commercially successful and economically important. If failure of an industry or of a product line is considered unacceptable, however, even without foreknowledge policy makers can pursue a strategy of *erring on the side of caution*. One tactic for this is *redundancy*—encouraging more new trials than may actually be necessary to launch successful lines of endeavor.

Third, because new policies on new subjects rarely perform perfectly, political participants need guidance on how to craft policies that will be feasible to implement, easy to monitor, and flexible enough to modify. Among other considerations, this implies minimizing up-front investments that cannot be recovered or reworked easily when the program needs to be altered or terminated. The costs of policy strategies and tactics almost always should be central, since inexpensive means for guarding against risks are obviously easier to justify than expensive means. (See Weimer and Vining 1990, pp. 311–322 on adaptive implementation; and Collingridge 1983, on flexibility.) Of course, a certain amount of background information about the new technology and its anticipated social problems will be necessary to convince policy makers that there *are* any risks, and to explain the strategies that deserve consideration in thinking about averting those risks. But information per se should not be the primary task, for facts and problems do not translate in any very direct way into policy options. Such translation into options for coping with uncertainty is precisely what analyses ought to provide.

Few technology policy analyses address all of these aspects of coping with uncertainty, and some miss them all. Thus, the National Academy of Sciences was mandated by Congress to provide advice regarding the division of labor in an international program of research on the greenhouse effect. And NAS was asked in particular to figure out what kind of research would be most helpful to policy makers. Instead, after two years and several hundred thousand dollars, NAS came up with a traditional report reviewing what scientists know about the greenhouse effect, and providing a list of more than a hundred topics—with no priorities—that allegedly "needed" to be done (NAS 1983, discussed in Morone and Woodhouse 1986).

In a very different field, the Office of Technology Assessment was asked to examine what government could do about easing problems encountered by workers when manufacturing plants shut down. OTA produced a credible report reviewing the problem, but provided little real guidance. There was extended discussion of the problems that would be created for businesses if government regulations were enacted, but no strategies and tactics specifically targeted at coping with these or other risks and uncertainties.

Finally, the General Accounting Office's study of the Seawolf nuclear attack submarine (SSN-21) program does deal with uncertainty. As with other military programs believed to be needed urgently, construction of the SSN-21 began before its design was complete. Worse yet, six of the submarine's ten *subsystems* are also being designed and built concurrently; and some of those subsystems will not be operationally tested *even as subsystems* prior to production. Failing to wait for thorough prototyping and testing is a prescription for difficulty, and GAO calls for "a less concurrent program," including a year's delay in awarding the next production contract (GAO 1990b). Unfortunately, while GAO dealt sensibly with the particulars, the agency's report did nothing to help legislators *conceptualize* more systematically what the options are for dealing with uncertainties of the types that routinely turn up in military procurement. So there is not likely to be much carryover from one analysis to the next, either for the analysts or for the users.

Partisanship

In addition to factual uncertainties, interpretations of risks require controversial value judgments, on which disagreement is certain. Should a new hazardous waste disposal facility be sited far away from cities, for example, to minimize the number of people potentially exposed if problems occur? Or should it be located near cities, so those who benefit from activities generating waste also bear a fair share of the risks? Should risk regulations aim to fully protect the most vulnerable members of the public, or should they aim to protect the average person? How much weight should be given to professional expertise compared with citizen perceptions (Walker, et al., 1983)? Several thoughtful, logical opinions can be given to each of these questions, none of which is authoritatively conclusive (Hiskes and Hiskes 1986; Johnson 1985; Shrader-Frechette 1983).

Again, every thoughtful analyst knows all this, and many reports point it out explicitly. How to adapt analysis to the inevitability of partisan choice is not a subject on which either policy analysis or social science more generally has had a great deal to say. One tactic is to steer clear of "partisanship" insofar as possible, and OTA avowedly endeavors to do. Unfortunately, it is impossible to be entirely nonpartisan: all "facts" are theory-laden, all theories are socially constructed in part, and all social constructions embody innumerable biases. In practice, therefore, nonpartisanship comes to mean some combination of:

- Avoiding alignment with one political party;
- Making assumptions that are widely accepted in the culture;
- Ducking issues that are too controversial;
- Acknowledging two or more sides of many issues.

Opinions differ as to whether this approach is more a pragmatic adaptation to political reality, or more a sham that naively disguises the inevitability of rampant partisanship in complex social analysis. But on either interpretation it seems fair to say that attempted-nonpartisanship does not necessarily do a good job helping political participants make *their* partisan judgments. It does not ask, "How can we help individual users of this report to balance competing values in a way that will work for them and their constituents?" Nor does it offer systematic assistance to the collective process through which tradeoffs are negotiated among large numbers of participants.

If individuals already know perfectly well how to make tradeoffs and don't need any help, and if political processes more or less automatically take care of such weighing for the collectivity, then technology policy analysts perhaps can legitimately set the issue aside. But while individuals and collectivities certainly have ways of weighing conflicting values, does anyone seriously allege that they do it nearly as well as would be desirable? (Compare Braybrooke and Lindblom 1963.)

If not, then we come back to the task of orienting analysis to assist in facilitating partisan judgment when values conflict. The easiest option is partisan analysis: adopt the viewpoint of one set of partisans, and explicate for them what would be a sensible reconciliation of their needs. This is actually what most technology policy analyses now try to do, albeit without owning up to it. The "set" of partisans is a large and diverse one, however, roughly equivalent to the middle seventy or eighty percent of the American political spectrum. This sort of partisan analysis is usually not very useful, because each aspect of a study is intended to be pleasing to a large majority of would-be readers/users. The analysis therefore has to proceed in terms of the lowest common denominator, and can rarely be very helpful to any particular user.

An alternative, rarely attempted but easy to envision, is multipartisan analysis. Address each significant major set of readers, in turn, clarifying the strategies each might wish to pursue. Then examine the prospects for bargained collective outcomes that would protect the concerns considered most vital by each set of partisans. Or, since it often will be impossible for analysts to guess where negotiations might eventually lead, at least distill and focus the partisan disagreements on decision axes, along which political interaction can take place.

In premanufacture screening of new chemicals, for example, there is a tradeoff: on one hand, many partisans want to encourage rapid innovation to compete effectively in global markets; but on the other hand, many partisans want to prevent introduction of potentially dangerous new chemical compounds. One obvious compromise is to stratify burden-of-proof requirements, as by requiring less testing for new chemicals that will replace older ones believed to be excessively dangerous. The analysis would concentrate on different ways that categories might be defined and on various aspects of burden of proof, clarifying for several sets of partisans what the implications of various approaches would be for their values, and identifying the lines along which political debate could profitably be directed.

There is a bit of this approach scattered here and there in some technology policy analyses, but it is not a standard part of our work.

Topic Choice

Professional social analysts receive elaborate training in the *execution* of projects, but very little in project choice. Whereas there is a voluminous professional literature on research methods, there is but a tiny trace of serious work on project choice (Lindblom and Cohen 1979; Lindblom 1990). Policy analysis texts do not supply much in the way of an explicit methodology for choosing topics for research, yet there are nearly an infinite number of possible topics and very limited budgets, misallocation of scarce analytic resources is virtually inevitable.

One way to narrow the search is to orient analytic priorities toward the needs of some group of potential users. Since most federal analytic efforts are partially "captive" to either executive or congressional patrons, analytic priorities almost automatically are set in part by the perceived needs of political partisans. That is widely perceived as a problem by observers outside Washington, and it certainly is troubling since less affluent and less organized citizens and groups rarely have access to the analysis they need.

But priority setting has to occur in some way, and no one knows enough to be sure what the highest priorities ought to be. Even if we somehow could rank problems according to the number of people potentially harmed, degree of harm, or some similar criterion, that would not translate into priorities for analytic attention. This is true in part because analytic attention ought to be targeted where it will make a critical contribution, and there are many significant problems where analysis will not do so. (For more detailed discussion, see Lindblom and Cohen 1979; Lindblom and Woodhouse 1993.) So bounding analytic attention by reference to the needs of some group of users makes excellent sense.

If usability is the key, moreover, there is a great deal to be said for letting influential potential users participate in shaping the agenda for analysis. They are far more likely to pay attention to analytic work that they've commissioned than work that comes to them cold. As every sensible analyst recognizes, however, potential users do not always know what they want, much less what they need. So some considerable massaging of their inputs is necessary, and an agency needs some basis for this task.

Members of Congress have been introducing bills on the subject of plant closings for over a decade, for example, but the legislation has gone nowhere because of business opposition. When asked to study the problem, then, an analytic team must ask itself precisely what sort of analytic contribution really is needed in such a case. Is it likely, after years of massive layoffs in the rustbelt, that the chief policy obstacle resides in legislators who still don't know that workers and their communities are harmed when manufacturing plants shut down? Or that a pre-closing notification law would be inconvenient and expensive for business? Not likely. If not, then fuller information on these matters may not be very helpful.

In contrast, it could be very useful for legislators and interest groups to learn more about the kinds of concessions to business that could evoke industry acquiescence to a notification law. Given its experience in serving as a broker among different stakeholders, OTA might well have been able to make a real contribution to this policy area if the agency had a better way of thinking about project choice in strategic terms: Who most needs to know what in order for there to be effective action on this problem? Consider another recent study, "Oil Production in Arctic National Wildlife Refuge," which examines experience to date with North Slope oil terrains and technologies as a way of gauging what might happen in ANWR. OTA found that the industry has learned enough to probably do a decent job of exploring ANWR without massive ecological damage, but notes a number of concerns. The study supports ANWR-exploration advocates' claim that Alaskan oil production declines in the next decade are highly probable, but they note major discrepancies among observers about the probability of actually finding oil in ANWR. Estimates range from the Department of Interior's 19 percent to some private estimates as high as 99 percent.

So far, so good. A policy maker who started out knowing nothing about ANWR would surely find the study helpful in providing background information, and in providing cues as to the credibility of the affected partisans' factual claims. Unfortunately, she or he would not necessarily be in a much improved position with regard to the crucial question of whether to approve the Department of Interior's plan for leasing oil exploration rights. Various partisans would have differing crucial questions on which their judgments might turn. Anyone willing to lease providing the oil companies had displayed competence in previous Alaskan ventures would be able to vote "yes" on the basis of the OTA study. So also anyone willing to lease if Alaskan oil production declines are upcoming. These people might have lacked crucial information provided by OTA.

But how many members of Congress could resolve the issue so easily? Not many, would be a good guess. One plausible point of view, for example, would hold that the vital question is: "Under what circumstances should I vote for exploration even if I'm worried about ANWR's environmental integrity?" The answer presumably would be: "When there is a sufficiently pressing need"—a matter that requires examining probable effects on energy prices and supplies in the absence of ANWR exploration (which the study does not attempt).

Others might pose their crucial question as, "When would ANWR oil make the most significant contribution to public welfare?" As with the previous question, this would require analytic attention to broader, longer-term energy issues than those considered in the ANWR study. Some of the relevant information was included in the report, such as the fact that there is pretty good agreement that the amount in question is not huge: on the order of about 3 billion barrels, which is just half a year's supply at current rates of consumption in the U.S. But these simple facts were not translated so as to be applicable directly to legislators' crucial questions. For example, OTA could have raised the possibility of saving the hypothetical ANWR reserves for possible future use in the middle of the next century, when worldwide oil supplies will be scarcer and higher priced.

By no means do I suggest that this conclusion is the one "correct" policy option that OTA should advocate. The point here is merely that analysts need to help policy makers figure out what their crucial questions ought to be, and, where possible, ought to help answer those questions.

Avoiding Inferior Topics

To save scarce time and attention for tasks like those outlined above, policy analysts must somehow avoid taking on inferior projects urged on them by superiors or by elected officials. This is no easy task, and requires better craft skills at project selection and defense than are now taught in policy analysis programs—or are developed on-the-job. Consider the announced rationale for the OTA background paper, "Public Perceptions of Biotechnology":

> Because government represents all of the public, it cannot ignore the concerns and preferences—no matter the extent of the misconceptions or how transitory the opinions might be—of any portion. It is important for policy makers to know not only what public opinion is, but also on what it is based (OTA May 1987, 3).

Does that imply a need for a study costing perhaps a hundred thousand dollars from a tight OTA budget? Hardly, for members of Congress are specialists at discerning the moods of their publics. Their antennae collectively are far better at tracking changes in the direction and intensity of opinion than even the best conceivable one-time polling (and the OTA/Harris poll was not state-of-the-art). While it is conceivable that high-level OTA decision makers had compelling arguments that made the study a top priority, no such rationale was advanced in print. And discussions with OTA staff suggest that their high level of good common sense does not equip them for skilled priority setting.

A related dictum can be derived from the previous section's discussion of coping with uncertainty. We found there that it is essential for analysts to ask, "What are the most serious risks against which some coping strategy is required?" To put it differently, analysis should be directed at assisting in the deployment of strategy. Why does that help with project choice? It becomes possible to rule out certain topics as high-priority candidates for analytic attention—namely those for which no influential political actor or analyst can propose a risk they find alarming. And after initiating preliminary inquiries into some alleged risks, analysts may terminate an inquiry because the alleged risk evaporates on closer inspection: if no plausible risk, then no need for a full study. Studies likewise can be terminated early when no plausible strategies and tactics can be elucidated for coping with the risk. For detailed analysis in support of strategy makes no sense without strategies being under active consideration.

What Type of Guidance for Users of Analysis?

More generally, we need some criterion for judging when a study can be terminated because it now provides "enough" help for potential users. If information is the goal, then presumably more information is always better than less. In principle, therefore, studies could go on indefinitely were they not terminated by arbitrary deadlines and budgets. While that method does work to a degree, it means that often studies which could be terminated early continue on, and those that ought to be extended are not.

More importantly, the study team lacks guidance on what they ought to be trying to provide for the would-be users. Consider the OTA report, "Commercializing High-Temperature SuperConductivity," which argues that traditional approaches to technology commercialization are inadequate to meet aggressive foreign competition. Nor has indirect commercial stimulation, as through tax policy, helped narrow the commercialization gap. Twenty policy options are analyzed, encompassing three distinct commercialization strategies, with the pros and cons of each responsibly summarized.

By conventional standards, it is a solid piece of work; and by any standard, it is a useful introduction to the field. Unfortunately, policy makers who did not already have a strategy in mind still would not know what to do, or even how to approach figuring out what to do. For the "two-handed" treatment of so many different policy options more easily confuses than enlightens. So the study is not really done: it is up to the policy analyst to massage a problem until an approach is developed to facilitate decision making by the partisans. That is not the same as recommending a single course of action, of course (see section on partisanship).

Another incomplete study was OTA's work on "New Developments in Biotechnology, Volume 1: Ownership of Human Tissues and Cells" (OTA March 1987). It provides a good illustration of the great difficulty analysts have in precisely targeting their inquiries. OTA examined the following questions:

- Are bodily substances "property" to be disposed of by any means one chooses, including donation or sale?
- Do individuals hold property rights to their genetic identity or do these rights adhere to the human species?

- Who owns a cell line—the person who was the human source of the tissues and cells, or the scientist who developed the cell line?
- Should disclosure, informed consent, and regulatory requirements be modified to cope with the new questions raised by the increased use of human biological materials? (OTA March 1987)

Those seem like interesting and important questions, integrally connected with the unfolding developments in genetic technology, clearly raising serious questions for policy making. So what's wrong? On closer inspection, it becomes apparent that the questions are not answerable by policy analysis, only by collective debate, will, and choice. (Selected aspects of these questions could be made analytically tractable, of course.) It is not uncommon to find social analysts taking on tasks that are well beyond their competence or right, so I do not wish to criticize OTA or technology analysts in particular. But considering how limited are the resources for analysis of social aspects of technology, it is especially unfortunate to find little craft knowledge of what *not* to study.

The OTA project team apparently came to this recognition by the end of their study, because they reframed the inquiry into much narrower policy issues. For example, should government explicitly permit commercial transfers of human tissues and cells, by mandating compensation, or regulating the conditions of market transfers? Alternatively, should buying and selling be prohibited, as under the National Organ Transplant Act? These still are not *answerable* via analysis, but at least they are more focused. Is it possible that the OTA team did not realize that such mundane questions could be addressed satisfactorily without reference to grander ethical questions which are unanswerable? In any event, they squandered much of a long report on the unanswerable before turning to explicit policy issues—and, partly as a result, provided inadequate guidance regarding the policy issues.

Combatting Naive Views About Technology

It seems incredibly elementary, but another task that must be achieved in technology policy analysis is helping political participants overcome their naive notions about "technology." There is a tendency to treat technologies as electro-, mechanical, or chemical phenomena which are neutral tools to be used—for better or worse—according to human choices. There is wide agreement among scholars who study technologies-in-society that such a tool-use viewpoint is much too limited. In application, any complex technology will be deployed by complex organizations, and it will enter into social life in intricate and often far-reaching ways (Bijker et al. 1987; Noble 1984; Winner 1977).

Even something as simple as a telephone ceases to be merely plastic, bell, electrical energy, and so forth when deployed as a system among users. It is a social instrument which cannot be understood apart from the meanings ascribed to it by the culture, the network of supporting organizations that service it, and the web of dependencies that get established around it.

Sensible technology policy would need to cope with a broad range of issues raised by changes in the sociotechnical phenomena we clumsily refer to as "technologies;" and their purely technical side arguably is their least problematic feature. It is by no means clear exactly what would be different about federal analytic efforts under a revised paradigm, but too many contemporary studies reinforce simplistic notions about technology that have been thoroughly discredited. Among other tasks, more adequate analyses somehow would call into question the outdated ideas.

La Porte has begun to chart directions by which this might be accomplished. He argues for analyzing technical phenomena so that "straightforward connections can be made between the proposed technology and the likely changes that the public, governmental agencies, and industrial organizations would experience" (La Porte 1984, 2). But this is rarely done: academic social scientists have not provided the requisite terminologies and conceptual frameworks; federal analytic efforts in technology policy tend to be directed by persons trained in the physical sciences who themselves have overly mechanistic ideas about the subject; and American political ideology turns attention away from intangible public goods, throwing responsibility for coping off the governmental agenda and onto individuals. So there will be limits to what analysts can get away with, but it should be possible at least to avoid depicting technologies as independent variables that "have impacts" on dependent social variables; for this implicitly separates something hardwarelike and causal from something social that is acted upon.

Rather than defining the technical and social separately, an alternative is to co-define them. Thus, technological developments would be seen, first and foremost, as a system of social and organizational relationships:

> The interactions between the organizations that help to realize the potential of technology-as-concept, the communities which are directly in contact with the technology, and the overarching societal institutions—legal, political, economic, and social—within which both the communities and the technologies-as-organization operate (La Porte 1984, 5).

Federal agencies now fairly uniformly take "technology" to be hardwarelike (or chemical, or . . .). While many studies are not oblivious to issues of scale, complexity, organizational routines, and other social considerations, neither are such issues at the core of their analyses.

In the myriad studies done on civilian nuclear power, for example, none has made central the simple fact that humans do not know how to successfully manage systems of the complexity of the one for which the Nuclear Regulatory Commission is responsible. Even a single nuclear-powered generating plant poses an insuperable challenge to all but a few of the most skilled utility companies. Managing a *system* of one hundred such plants, each with thousands of employees, is beyond current human capacity. This being the case, the pressing policy issue is how to learn to make the regulatory task feasible—a fascinating and important issue about which policy makers presumably would be glad for advice. But they never have gotten it in any serious detail.

The same shortcoming afflicts studies of defense procurement. An emphasis on the technical (including cost effectiveness) obscures key social aspects which would have to be coped with for policy to be successful. Demchak's research on the M-1 tank, for instance, showed how "the Army embarked on its modernization with a poor understanding of the (organizational) consequences of introducing such complex weapon systems." Because complex, high-technology tanks don't function without "expensive, multi-layered printed circuit boards, . . . when those tanks are damaged, someone will have to be there who can repair such machines fast and accurately because there are so few of them. And that someone is likely to be deep in the rear of the battlefield, unwilling to go forward and risk a truck of needed parts or scarce human expertise since both are in short supply" (Demchak 1988; Demchak 1987).

The M-1 Abrams tank was developed during the late 1970s, and its problems have become well known. So it would be reasonable to suppose that the lessons would by now be integrated both into relevant policy analysis and into army R&D decision processes. The evidence suggests room for doubt. While there certainly has been growing concern about the performance dysfunctions and cost overruns of complex weapons systems, little *conceptual* improvement is discernible. Thus, the Apache helicopter suffers from frequent component failures and performs poorly in less-than-optimal weather; instead of a learning curve, GAO found that the Apache's availability rates actually *decrease* as battalions accumulate flight hours. While this should have been predictable from previous experience with technologically complex and finicky weaponry, the Army still is providing undersized maintenance organizations and weak repair capabilities not unlike those that have characterized the M-1 tank (GAO 1990).

This sort of organizational perspective on technology is not wholly absent from contemporary studies, but rarely does it take the central position that scholars who have studied technologies in-application believe it deserves.

Conclusion

In sum, political actors from presidents to community organizers face high uncertainty, rampant disagreement with others in and out of their working group, and sharp tradeoffs among competing values. They have inadequate time, energy, and resources to digest more than a fraction of the information potentially available to them. And they often do not know how to conceptualize the problems in their domain of responsibility, or how to think strategically about them. Good technology policy analysts can provide more usable knowledge if we explicitly set out to cope with these actual conditions facing potential users of our work.

All of us know about these constraints on the use of information in political life. But we are steeped in western cultural traditions that grossly exaggerate the possibility of proceeding through life via understanding, and that tend to equate understanding with semi-objective knowledge which is more or less added up, one bit upon another. Hence, all of us fall back at times toward a neutral-understanding approach to analysis, believing against our own better judgment that more information per se will lead to a better world.

I have tried to suggest that it may be possible to partly counteract this insidious habit of thought. To do it will require development and use of countervailing heuristics to help identify key risks in various policy areas, as well as to develop a repertoire of strategies and tactics for protecting against those risks. There will be limits to how far we can go anytime soon, limits placed upon us partly by our audiences who themselves are steeped in naive notions about the role of knowledge in decision making. But if we have clearer and more powerful ways of targeting problem selection and project execution, we will be in a position to negotiate as effectively as possible with those whom we seek to assist.

Notes

1. This includes *unforeseeable* consequences: while much harder to guard against, there are tactics available. For further discussion, see Woodhouse and Hamlett, 1992; Morone and Woodhouse, 1986.

Bibliography

Ascher, William. 1978. *Forecasting: An Analysis for Policy-Makers and Planners.* Baltimore: Johns Hopkins University Press.

Bijker, Wiebe E., Thomas Parke Hughes, and Trevor Pinch. 1987. *The Social Construction of Technological Systems: New Directions in the Sociology and History of Technology*. Cambridge: MIT Press.

Collingridge, David. 1980. *The Social Control of Technology*. New York: St. Martin's.

————. 1983. *Technology in the Policy Process: Controlling Nuclear Power*. New York: St. Martin's.

Collingridge, David, and Colin Reeve. 1986. *Science Speaks to Power: The Role of Experts in Policy Making*. London: Frances Pinter Pubs.

Demchak, Chris C. 1987. *War, Technological Complexity and the U.S. Army*. Ph.D. diss., Department of Political Science, University of California, Berkeley.

————. 1988. "Complexity, Rogue Outcomes and Weapon Systems," presented at the 1988 annual meetings of the American Political Science Association, Washington, D.C.

Drucker, Peter. 1986. "New Technology: Predicting Its Impact." In *Technology and the Future*. 4th ed., edited by Albert H. Teich, 214–218. New York: St. Martin's.

Falk, Jim, and Andrew Brownlow. 1989. *The Greenhouse Challenge: What's to be Done?* New York: Penguin.

Hiskes, Anne L., and Richard P. Hiskes. 1986. *Science, Technology, and Policy Decisions*. Boulder: Westview.

Johnson, Deborah. 1985. *Computer Ethics*. Englewood Cliffs, NJ: Prentice-Hall.

LaPorte, Todd R. 1984. "Technology as Social Organization." Institute of Governmental Studies: University of California at Berkeley.

Lindblom, Charles E., and David K. Cohen. 1979. *Usable Knowledge: Social Science and Social Problem Solving*. New Haven: Yale University Press.

Lindblom, Charles E. 1990. *Inquiry and Change: The Troubled Attempt to Understand and Shape Society*. New Haven: Yale University Press.

Lindblom, Charles E., and Edward J. Woodhouse. 1993. *The Policy-Making Process*. 3rd ed. Englewood Cliffs, NJ: Prentice-Hall.

Lowe, Ian. 1989. *Living in the Greenhouse*. Newham, Australia: Scribe.

Morone, Joseph, and Edward J. Woodhouse. 1986. *Averting Catastrophe: Strategies for Regulating Risky Technologies*. Berkeley: University of California Press.

————. 1989. *The Demise of Nuclear Energy? Lessons for Democratic Control of Technology*. New Haven: Yale University Press.

National Research Council. Carbon Dioxide Assessment Committee. 1983. *Changing Climate*. Washington, D.C.: National Academy Press.

Noble, David. 1984. *Forces of Production: A Social History of Industrial Automation*. New York: Alfred A. Knopf.

Shrader-Frechette, Kristin S. 1983. *Nuclear Power and Public Policy: The Social and Ethical Problems of Fission Technology*. 2nd ed. Boston: D. Reidel.

Stross, Randall. 1989. *Technology and Society in Twentieth Century America*. Chicago: Dorsey.

U.S. Congress. Office of Technology Assessment. 1982. *Genetic Technology: A New Frontier*. Boulder: Westview.

————. 1984. "Nuclear Power in an Age of Uncertainty." Washington, D.C.: GPO.

————. June 1988. *Commercializing High-Temperature Superconductivity*. Washington, D.C.: GPO.

U.S. General Accounting Office. 28 September 1990. "Apache Helicopter: Serious Logistical Support Problems Must Be Solved to Realize Combat Potential." GAO/NSIAD-90-294.

———. 28 September 1990, "Navy Ships: Concurrency Within the SSN-21 Program," GAO/NSIAD-90-287.

Walker, Charles A. et al. 1983. "Value Issues in Radioactive Waste Management." In *Too Hot to Handle: Social and Policy Issues in the Management of Radioactive Wastes,* edited by Charles A. Walker et al., 184–206. New Haven: Yale University Press.

Weimer, David L., and Aidan R. Vining. 1989. *Policy Analysis: Concepts and Practice.* Englewood Cliffs, NJ: Prentice-Hall.

Wildavsky, Aaron. September 1983. "The Assessment of Safety Goals and Achievements in Complex Technological Systems: The Integration of Technological and Institutional Considerations." Unpublished manuscript prepared for the U.S. Nuclear Regulatory Commission.

Winner, Langdon. 1977. *Autonomous Technology: Technics-out-of-Control as a Theme in Political Thought.* Cambridge: MIT Press.

Woodhouse, E. J. 1988. "Sophisticated Trial and Error in Decision Making About Risk." In *Technology and Politics,* edited by Michael Kraft and Norman Vig, 208–223. Durham: Duke University Press.

Woodhouse, E. J., and Patrick Hamlett. 1992. "Decision Making about Biotechnology." In *The Social Response to Environmental Risk,* edited by Daniel W. Bromley and Kathy Segerson. Kluwer Academic.

3

Institutional Structures and Technology Policy in Congress: Toward an Applied Policy Science

William W. Keller

Can a purposefully divided government with co-equal branches make consolidated policy—especially policy with industrial or technological implications? The short answer would seem to be no. Discussion of industrial policy is barely tolerated in political and business circles in the United States. And while industry, the military, and universities lobby for more government-sponsored research and development funding, the actual coordination of civilian R&D programs is left to the private sector. Government is unable or unwilling to pick winners and losers; these must be determined by more impersonal forces in the international economy. Congress remains inhospitable to industrial or comprehensive technology policy, even as the U.S. piles up negative trade balances and loses technological advantage to nations that unabashedly target industrial sectors and take over foreign markets. Perhaps it is part of the American ethic to carry ideology beyond its logical and practical limits. Perhaps inability to act in the face of hegemonic decline is embedded in the Constitutional structure of the polity. For whatever reasons, it is indeed difficult to see how a system so pluralistic and riddled with factions can compete in the international marketplace without adopting superior forms of technological and industrial organization.

This dilemma is founded on the truism that politics, not policy outcomes or analysis, dominates the agenda in Congress. While it is still in its infancy, the art of policy analysis—and the institutions Congress has created to foster it—can and increasingly does affect the nature of the policy process. Stated more formally, the character of the policy process changes with the institutions that produce the process. As Congress institutionalizes policy analysis, such analysis will influence the outcome of its deliberations. The process will be more or less political, depending on the formal and legal attributes of the institutions that conduct policy analysis for the Congress.

Congress was slow to understand the building up of the modern liberal state and slower still to equip itself with the staffs and analytical resources needed to legislate in a technological age.[1] A highly centralized committee structure, revolving around a few powerful members, lacked the versatility and specialization necessary to check or balance multiple centers of concentrated executive power. Until the rise of subcommittee government,[2] Congress depended on information and data provided by the executive agencies and by a host of special interests. There is no reason to expect that an entrenched executive bureaucracy will provide balanced and unbiased information and analysis which could be used to undermine an Administration's position. And there is every reason to assume that lobbyists will take active steps to promote their own agendas and investments. If these arrangements created comfortable iron triangle relationships, they ultimately resulted in a federal government characterized by interest group dominance of the policy process, and a state structure less legitimate by Madisonian principles because it ultimately lost its capacity to plan effectively and to achieve justice among a multitude of competing interests. (Lowi 1979). As post–New Deal government learned to accommodate special interests, the public interest became less important as an electoral factor, and was simply assumed. Aggregate interests may be far more extensive and important than the sum of privileged, special interests. The problem with pluralist theory is that it is essentially a political reconstruction of Adam Smith's economic theory of the invisible hand. For reasons that were never adequately explained, pluralists believed that the aggregate of political pressures emanating from organized interests would add up to the public good. Public policy analysis must therefore begin with a rejection of the pluralist vision. But this is not enough. It must also be insulated[3] from political forces that would require it to advocate partisan and other political pressures.

It was necessary for Congress to reorganize and to equip itself with an analytic capability in the early 1970s, if for no other reason than to stay on a more or less equal footing with the executive power house it had created. This is not to suggest that congressional politics has somehow been replaced by a more systematic method of generating public policy, but only that the outcomes of politics can be informed and improved, and that Congress has already experimented with various institutional structures, and with varied success.

Congress took a number of steps in the early and middle 1970s to establish institutions for policy analysis and to place them under direct congressional control.[4] It converted the old Legislative Reference Service into the Congressional Research Service (CRS), greatly expanding its staff and resources. Congress also expanded the General Accounting Office significantly, providing for members the ability and staff to peer into the operations of the executive, to reach far beyond the traditional oversight powers of the investigative committees. In addition, Congress set up the Office of Technology Assessment (OTA) in 1972 to provide non-partisan, balanced analysis of complex technological matters. In 1974, the congressional Budget Office (CBO) was established, enhancing Congress's ability to challenge the executive budget, and to forecast the fiscal consequences of various legislative options. If knowledge is power, Congress understood the latter and set about equipping itself with the former. The departure was that Congress constituted new parts of itself, expanding its analytical resources, and creating the means to inform the debate with information not generated to serve the purposes of the executive branch or of special interests.

While the mainstream congressional literature tends not to focus on technology, the received wisdom is that Congress, for a very large number of reasons, cannot make disinterested, intelligent technology policy that operates in the public interest.[5] Members are too concerned with constituency service and with positioning themselves for the next election (Mayhew 1974, 16–17; Fiorina 1977, 37). Committee jurisdictions are too complex or too narrow or too Byzantine to deal with issues that are economy-wide in scope (Davidson 1976, 300–1). There are too many entrenched interests, too many political action committees with too many agendas and too much political power to be resisted. There is grave doubt in the literature regarding the ability and political will of Congress to place national objectives on an equal footing with distributive politics and other parochial concerns associated with electoral survival.

In addition, the term "technology policy"is used to refer to two different processes and congressional environments. The first is the large scale, high-tech projects, such as the Apollo moon shot and the superconducting-supercollider, that are sometimes thought of as the Nation's science and technology agenda. These kinds of items are often, but not always, handled by the science and technology committees of Congress. This kind of S&T policy is almost entirely distributive in character, and a list of such projects over the years would provide excellent case studies in the art of congressional horse-trading and pork-barreling.

But this is only the tip of the iceberg of technology policy. A second kind is harder to define because it extends across the legislative and economic landscape. It consists of a wide array of policies that depend on an understanding of the environmental, economic, and societal implications of complex technologies. It is eclectic and ad hoc in nature because technology does not respect committee jurisdictions, and plays an important role in hundreds of policy areas. It may even be decisive. For example, advanced electronics technology plays a central role in the overall military strategy of the United States. It is also critically important to the continued survival of many industrial sectors in the civilian economy. But tax, procurement, and education policies structured to enhance defense electronics might involve economy-wide changes with broad implications for civilian industry. Technologically intensive policy debates tend to generate all stripes of experts on all sides of the question, and technology issues tend not to divide along partisan or ideological lines. The level of discourse cannot easily be condensed into briefings that members can absorb quickly[6] or be translated into twenty second sound bites for mass media consumption.

Despite its broad application to a multiplicity of policy problems, technology, per se, is not a dominant area of congressional concern. The science and technology committees of Congress are decidedly second tier in terms of seniority of members, relative power within Congress, and allocation of budget authority. Nevertheless, understanding of technology and technological considerations are increasingly important in the day to day business of the Finance, Ways and Means, Defense, and Commerce committees, among others. Congress will continue to construct policy in scores of areas where technology plays a central role, and Congress will act to fund big science and technology projects, and to support basic and applied research, in both the military and civilian sectors. My argument is that contrary to majority opinion, Congress can legislate effectively in areas that require technological sophistication, and in areas where technology is decisive to the outcome of the issue. The rudimentary instruments are already in place; what remains to be seen is whether they acquire the requisite institutional force to alter the character of the policy process itself. ·

In his famous essay on Congress, Huntington concluded that Congress should recognize that its legislative function was no longer primary, and focus instead on oversight and constituency service activities. Congress could represent the interests of the electorate within the bureaucracy, even if it could not exercise its constitutional responsibilities. In this way, the legislative branch could preserve itself as an independent institution. The alternative, according to the essay, would be sustained institutional stalemate and a continuous risk of a constitutional crisis (Huntington 1973, 33, 37–8).

What Huntington anticipated was a legislature that works at the margins of public policy, ratifying, with minor alterations in budget and emphasis, the program that is proposed by the president. Many of the real policy decisions are indeed contained in the budget submissions of the executive agencies, and many of these do make their way through the congressional process unscathed. But increasingly often, Congress challenges the agenda of the executive, forcing modifications based on its own independent analysis. Huntington's prescription was a formula for a vastly weaker legislature that did not materialize. Its central premise that Congress had failed to adapt to changes in society was premature. It failed to recognize that Congress is capable of reconstituting itself in response to new forms of social, technological, and political organization.

The question of whether Congress can in fact make intelligent and coherent technology policy is bound up in the distinction between policy and politics. Politics has more to do with process, policy with substance. Politics is informal and personal, policy has the force of law. Politics is urgent and focussed on the immediate need, policy conditions the future and is permanent until it is reinterpreted or replaced. Policy is the rational, legal, and administrative outcome of a process which is itself unpredictable and disorderly. From the congressional perspective, the policy that is enunciated is contained in the language of the public law. And yet it is possible to create the political and institutional space for independent analysis in the relationship between politics and policy.

To understand the place of policy analysis in Congress, it is useful to abstract from the political context precisely because politics is hostile to independent and balanced analysis. As politics is transformed into policy, the potential at least exists in Congress to situate policy analysis in a neutral zone some place between electorally motivated political behavior and formal legal outcomes of the policy process. But to do so, the policy analysis function must be sufficiently insulated from politics. That means that it must be institutionalized in particular ways. The structure and operations of the Office of Technology Assessment, as contrasted with that of other congressional agencies, provides some important insights.

When Congress passed the Technology Assessment Act of 1972, it not only created an institutional capacity for independent analysis of technology policy, it also designed processes to protect OTA from political pressure. It did this by creating a number of buffers between the new office and its congressional clients. First, by statute, Congress placed the resources of OTA beyond the immediate grasp of most members and their staffs. Typically, a study can only be initiated at the request of the chair and ranking minority member of one or more committees.[7] This distinguishes OTA from the other congressional agencies that respond directly to the requests of individual members.[8] For this reason, requests handled by the CRS, the GAO and, to some extent, the CBO are more likely to be responsive to the political and electoral concerns of members.

A second institutional buffer resides in the composition and functions of the Technology Assessment Board (TAB) that oversees and directs OTA's operations. When OTA receives a request for an assessment from a committee of Congress, it does not respond directly to the committee. Instead, the agency, when appropriate, prepares a proposal for an assessment—which is responsive to the committee's request—but which may address additional considerations. The proposal is then taken before the TAB, and it is approved or disapproved by a vote of the Board. In this way, the OTA policy analysis function is twice removed from electoral and other political motivations of individual members of Congress.

The distance from partisan and bicameral politics is also institutionalized in the composition of the OTA Board, which is an unusual committee, especially by congressional standards. By law, the TAB is composed of six senators and six members of the House. Three of the senators are selected from each party by the President pro tempore of the Senate, and three congressmen are selected from each party by the Speaker of the House.[9] The chair and vice chair are selected by the members of the TAB and alternate between each House and each party from one Congress to the next.[10] The net effect of these rules is to keep partisan politics out of the assessment process. Even Ted Lowi could not complain of the specificity of the OTA Act through which Congress delegated control of OTA to a group of its members, but also succeeded in creating an institutional structure that largely protects the agency from political motivations and electoral ambitions. A third buffer consists of the internal institutional structure of OTA, and the organization's policies and procedures that have been carefully designed to ensure a balanced and independent analysis of technology policies and issues.[11] Although the topic at hand does not permit an in depth analysis of OTA's assessment process, it is worth noting that each assessment is advised by a panel of distinguished experts drawn from industry, the scientific and technical communities, public interest groups, labor, academic circles, and from the ranks of retired government and military officials. These advisory panels are chosen so that they represent diverse interests and points of view. One measure of political independence is that it is possible to work at OTA as an analyst and encounter little or no political pressure from the Hill. For these reasons, capture of the agency by an entrenched interest would be extremely difficult; OTA would lose its credibility and there would be little, if anything, to gain.

But OTA is not so insulated from Congress that it cannot use the power of Congress to gather information and to bring together diverse and interested parties and stakeholders of an issue. In this respect it enjoys a tremendous advantage over the private sector think tanks such as the Brookings Institution and the American Enterprise Institute. OTA is authorized by statute to secure information directly from any executive agency, and the executive agencies are directed to "furnish the information, suggestions, estimates, statistics, and technical assistance directly to the Office [OTA] upon its request."[12] In actual practice, however, most of the executive agencies and a very wide range of private institutions and groups are eager to become involved in OTA's work and to provide information and analysis. When Congress established OTA and the other congressional agencies, it did so because it recognized the power of information and persuasive analysis, and wished to create its own sources that would be independent of vested interests and the executive branch of government. Congress intended that the analysis provided by the congressional agencies would inform and influence its members, enabling them to make sound judgments in very difficult and complex policy areas. In the space remaining, I can only suggest a number of ways in which the congressional agencies do influence the policy process.

The first has to do with the perennial problem of tangled or insufficient committee jurisdiction, where a committee identifies a policy problem, but does not have the specific knowledge or jurisdiction to address it. Or conversely, several committees may have overlapping jurisdiction, with the result that a series of different policy approaches are developed, using different information resources and with little or no coordination. In such cases, OTA can establish formal coordination of committee research efforts by conducting an assessment of a policy area at the request of several committees of jurisdiction. In the process of beginning the assessment, the policy concerns of the leadership of the different committees are often built into the design of an OTA study. In this way, a kind of formal coordination is implemented after it has been approved and receives the backing of the OTA Board.[13]

Second, the congressional agencies can influence Congress by helping to develop and inform the context in which a policy question is debated. But here, as before, institutional characteristics are extremely important. The agency must have the capacity to take the longer view not traditionally associated with politics, and the ability to study all relevant sides of a question. In some cases, the agency will inevitably be drawn into political disputes, even with the executive departments, as a result of its analysis. But in principle, and in the day to day working environment, the agency must be able to frame the issues independently of the political process, and to forecast or anticipate policy solutions before the issue at hand has become politically hot. In this way, the policy approach may exist before the fact of the politics, and may condition the way in which the politics develops.

Perhaps the most direct way in which an agency of Congress can influence public policy is when its analysis is used to support specific legislative language that becomes law. There are numerous examples in which the language of policy options developed by OTA has been codified by Congress. There are also many examples where individual members of Congress have cited OTA research, reports, and policy options to support their legislative proposals (OTA, 1987, 1988, 1989).

In conclusion, Congress equipped itself in the middle 1970s with analytic resources needed to inform and reform itself in complex areas where the technological and fiscal implications of policy are not intuitively obvious. In the process, it greatly enhanced its own powers to question the basic policy directions of the executive branch. In turn, the information agencies of the Congress have begun to influence the character of the policy debate and to make modest, yet detectable, changes in the nature and outcomes of the policy process itself. The degree of influence of policy analysis depends upon the institutional framework within which the analysis takes place. To the extent that they are independent of and sufficiently insulated from day to day politics, the congressional agencies offer the opportunity for the evolution of an applied policy science to which political scientists might make a substantial and constructive contribution. The role of the information agencies is important because it is not entirely clear how a government purposefully divided against itself—with a legislative branch empowered to say no to the executive and yes to special interests—can compete in an international environment of increasing technological, economic, and military parity and decreasing sovereignty.

Notes

1. It was not until the middle 1970s that the slow expansion of committee staffs initiated in the late 1960s really began to take off. For example, the House Science Committee staff expanded from approximately 30 in 1975 to about 90 in 1979. By 1979, the total of congressional staff had reached 23,528 (Malbin, 136).

2. For the purposes of this discussion, I refer to Davidson's (1988, 349–50) division of three different eras of the modern Congress which he called the "era of strong committee chairmen" (1937–1964), the "rise of subcommittee government" (1965–1978), and the "era of contraction" (1979–).

3. I have used the concept of insularity elsewhere to establish the degree to which the activities of an executive agency would remain unknown to the Congress and other units of government. Here insularity refers only to the ability of the congressional agencies to work without political interference. See William in Keller, *The Liberals and I. Edgar Hoover: Rise and Fall of a Domestic Intelligence State* (Princeton, NJ: Princeton University Press, 1989), pp. 19–23.

4. A principal finding of the OTA Act is that "the Federal agencies presently responsible directly to the Congress are not designed to provide the legislative

branch with adequate and timely information, independently developed, relating to the potential impact of technological applications." P.L. 92–484, sec. 2(c) (1).

5. Some observers believe that policy analysis will, of necessity, be captured and weakened by politics (Jones, 264). I take exception to this view below.

6. Fewer than twenty members of Congress hold degrees in scientific or technical fields.

7. The Act does provide other means of initiating a study, but because OTA receives many more requests than it can fill, these are rarely exercised. See P.L. 92–484, Sec. 3(d).

8. In this respect, the CBO occupies a middle ground between OTA on one side and CRS and GAO on the other. Although this is not specified in its organic legislation, by custom, CBO honors requests from the chairperson or ranking minority member of full committees, the chairperson of a subcommittee or task force, and the leadership of the Senate and the House of Representatives, Unlike OTA, it does not have a congressional board that directs its activities.

9. In fact, the minority leadership in each House recommends the minority members of the TAB to the Speaker and the President.

10. In practice, the chair and vice chair are always chosen in accordance with their seniority on the Technology Assessment Board.

11. Despite its carefully studied balance, OTA does make specific findings and presents policy options and alternatives to the Congress.

12. P.L. 92–484, Sec. 6(d). In addition, the Technology Assessment Board may issue subpoenas over the signature of the chairman. Sec. 4(d).

13. The efforts of the TAB in this regard are not inconsequential because its membership includes several of the most powerful members of Congress, including five full committee chairmen. The committees associated with these chairmen are major clients of OTA.

Bibliography

Brooks, Harvey. 1984. "The Resolution of Technically Intensive Public Policy Disputes." *Science, Technology & Human Values* 9(1): 39–50.

Davidson, Roger H. 1976. "Congressional Committees: The Toughest Customers." *Policy Analysis* 2: 299–323.

———. "The New Centralization of Capitol Hill." *Review of Politics* 50 (Summer): 345–64.

Fenno, Richard F., Jr. 1978. *Home Style: House Members in Their Districts.* Boston: Little, Brown.

———. 1986. "Adjusting to the U.S. Senate." In *Congress and Policy Change*, edited by Gerald C. Wright, Jr. and Leroy N. Rieselbach. New York: Agathon Press, 123–47.

Fiorina, Morris P. 1977. *Congress: Keystone of the Washington Establishment.* New Haven: Yale University Press.

Frye, Alton. 1976. "Congressional Politics and Policy Analysis: Bridging the Gap." *Policy Analysis* 2 (Spring): 265–281.

Gibbons, John H. 9 April 1981. "The Role of Analysis in Government." *Midcontinent Perspectives* Kansas City: Midwest Research Institute.

————. 1984. "Technology Assessment for the Congress." *The Bridge* 14 (Summer): 2–8.

Hamlett, Patrick W. 1983. "A Typology of Technological Policymaking in the U.S. Congress." *Science, Technology & Human Values* 8(2): 33–40.

Huntington, Samuel P. 1973. "Congressional Responses to the Twentieth Century." In *The Congress and America's Future*, 2nd ed., edited by David B. Truman. Englewood Cliffs, NJ: Prentice-Hall. Jones, Charles O. 1976. "Why Congress Can't Do Policy Analysis." *Analysis* 2 (Spring): 251–264.

Keller, William W. 1989. *The Liberals and J. Edgar Hoover: Rise and Fall of a Domestic Intelligence State.* Princeton, NJ: Princeton University Press.

Lowi, Theodore J. 1979. *The End of Liberalism.* 2nd ed. New York: W.W. Norton.

Malbin, Michael J. 1981. "Delegation, Deliberation, and the New Role of Congressional Staff." In *The New Congress*, edited by Thomas E. Mann and Norman J. Ornstein. Washington: American Enterprise Institute.

Mayhew, David R. 1974. *Congress: The Electoral Connection.* New Haven: Yale University Press.

Office of Technology Assessment. 15 December 1986. "Fiscal Year 1988 Justification of Estimates." Submitted to the Subcommittee on Legislative Branch Appropriations.

————. "Fiscal Year 1989 Justification of Estimates." 21 December 1987. Submitted to the Subcommittee on Legislative Branch Appropriations.

————. 16 December 1988. "Fiscal Year 1990 Justification of Estimates." Submitted to the Subcommittee on Legislative Branch Appropriations.

Price, David E. 1985. "Congressional Committees in the Policy Process." In *Congress Reconsidered*, edited by Lawrence C. Dodd and Bruce I. Oppenheimer. 161–188. Washington: Congressional Quarterly, Inc.

Sundquist, James L. 1981. *The Decline and Resurgence of Congress.* Washington: The Brookings Institution.

Thurber, James A. 1981. "The Evolving Role and Effectiveness of the Congressional Research Agencies." In *The House at Work*, edited by Joseph Cooper and G. Calvin Mackenzie. 292–315. Austin: University of Texas Press.

Whiteman, David. 1985. "The Fate of Policy Analysis in Congressional Decision Making: Three Types of Use in Committees." *Western Political Quarterly* 38 (June): 294–311.

PART TWO

Case Studies in Congress and Policy Analysis

4

Congress, the Environment, and Technology Assessment

Kathryn D. Wagner

Introduction

Science, politics, and economics determine the evolution of most environmental issues. The levels of certainty, contentiousness, and cost posed by an environmental problem combine to create the dynamic of the policy making efforts to address it. One of the most striking features of environmental legislation to many observers is its unrealistic, complex, and cumbersome nature. The process of policy articulation is complex and relatively unpredictable, but much can be learned from studying how the environmental policies have evolved to their current state of both intricacy and, in many cases, impasse.

Among the reasons for the failure of current environmental laws to meet their goals are an inadequate scientific understanding of the issue, technical difficulties to address it adequately, and constraints imposed by the statutory language. These failures, particularly the basic statutory design of an environmental law, appear to have determinative effects on the form and type of policy implementation (Harris and Milkis 1989; Bosso 1987; Wagner 1987). The statute itself determines the extent to which technical feasibility, costs, and risks should be considered in the policy making process.

As the brief examination of the contrasting responses to the ambient air pollution and medical wastes issues in this papers shows, the degree to which Congress factors such considerations into environmental legislation is both a product of the time in which the law was drafted and the degree of contentiousness over the policy issue. The comparison suggests that technology assessments and other technical studies by congressional agencies are important sources of scientific and technical information for Congress which are increasingly used throughout the policy articulation process. A broad, working definition of technology assessment is a focused effort to distinguish and clarify the technical and nontechnical (e.g., social, economic and political) aspects of an issue to facilitate development of policy options.

Technology assessment is similar to all policy analysis in that it seeks to elevate the level of public debate by: 1) clarifying the values and issues at stake in a policy dispute; 2) distinguishing competing policy goals; and 3) identifying viable policy options in light of these different policy goals. The purpose is to facilitate consideration of the "right questions" (i.e., those which address the complexity of the issue and are action-oriented) in the policy process. Ultimately, the structure of the policy process is key. Central to its definition is whether Congress assumes the responsibility when drafting legislation to make explicit choices among competing policy goals and options.

The importance of scientific information and technical limitations to resolving a policy problem and the critical issue of how Congress makes decisions despite scientific and technical uncertainty are well illustrated by the efforts to set national ambient air quality standards (NAAQSs). The Clean Air Act Amendments of 1970 (42 U.S.C.A. Sec. 7401 et seq.) directed the Environmental Protection Agency (EPA) to establish NAAQSs to protect human health and the environment. Despite revisions since its initial passage, the goal of eliminating air pollution remains elusive. The more recent legislative activity on the medical waste problem provides a marked contrast. In 1988, after a summer of beach wash-ups of medical waste caught the media's attention and raised public concern, Congress passed the Medical Waste Tracking Act (42 U.S.C.A. Sections 6901 et seq.).

In the medical waste policy area, Congress is more cautious in its assumptions of the certainty of scientific understanding of and technological solutions to the problem. In part this may be a reaction to the state of stalemate resulting from the ambitious legislative efforts of the early 1970s. It may also be that for newer areas of legislation, for which policy interests are not yet entrenched, there is greater ability or willingness for Congress to address an issue on nonpolitical grounds, (i.e., more on the basis of its scientific and technical aspects than on partisan concerns). Under such circumstances, the greater latitude in policy development for Congress also may be reflected in more carefully written statutes with more explicit definitions of the policy problem and directions for agency action.

The Challenges of Statutory Design and Scientific Uncertainty for Congressional Environmental Policy Making

Environmental laws, as a crucial component of contemporary politics, provide a lens through which the use and impact of science, technology and technical information on policy making can be studied. The factors largely responsible for the meager implementation results of the laws to date are the nature of the statutes themselves and the problem of scientific uncertainty. These characteristics are problematic for Congress as it attempts to initiate a new, major policy or to re-direct an existing law.

There is no dispute that when the environmental legislation was passed in the early 1970s, the American public was eager for dramatic change in the way the nation used and related to its resources (Mitchell 1984; Schoenbrod 1983).[1] The laws were written, however, with little consideration of their feasibility. For example, even whether national air quality standards *could* be achieved under the best of circumstances was not a primary concern at the time (Beam 1981; Schoenbrod 1983). It is hardly surprising that the agency actions flowing from the enabling legislation has been characterized as efforts "to implement beyond capability" (Jones 1975).

The passage of the Clean Air Act Amendments of 1970 represented not only the most comprehensive federal environmental legislation to date, but raised a whole host of implications for the political system beyond mere pollution abatement. Along with the passage of the Clean Water Act during this same time, this law was an attempt by Congress to correct in part for its abdication of power during the New Deal era. The majority-building incrementalism of the New Deal Congress was replaced by an agency-forcing strategy. Congress gave some administrative agencies, such as the Environmental Protection Agency, strict, perhaps overly-specific, deadlines and detailed procedures to follow. In other areas, the Agency was given a high level of discretionary power, yet often only to find itself narrowly constrained by other sections of the law.

Congress was certainly attempting to shift the organization of the power to determine policy back to itself. At the same time it was also attempting to prevent the "capture" of various agencies by special interests by curtailing their discretion. This strategy, as readily apparent from the implementation efforts under the Clean Air Act, worked ironically to prevent agency capture only to create agency crippling (Wagner 1987).

The consequences of this legislation have been significant, if sometimes unexpected. Yet, certain of the unexpected consequences of the laws may not have been avoided even under more ideal circumstances for legislative drafting. It is true that the laws have sometimes led to regulation of the least harmful pollutants (see below) and other policy misdirections which have proved difficult to overcome. Often enough, however, scientific understanding of an issue does not parallel legislative development of a policy.

Other unexpected outcomes of environmental legislation though are the result of the failure of mechanisms in the laws which in fact represented conscious efforts by the legislators to design the most effective statutes. The implications for modern policy making of these failures are far reaching. These aspects of statutory design have the potential for structuring the entire implementation process and determining the degree of flexibility and authority an agency will have to make necessary adjustments based on new scientific or technical information.

Statutory Design

Action-forcing, the main innovative mechanism of environmental laws, was designed to ensure effective agency action and speedy implementation. The two basic variations of action-forcing are: technology-forcing and agency-forcing. The former includes provisions in the laws which require an agency to set technology-based standards for pollution control creating incentives for technological innovation by industry. The latter includes provisions containing strict deadlines by which agency action must be taken. Most assessments of these techniques have concluded that they have largely failed to produce their intended twin results of curtailing agency discretion and ensuring rapid implementation (e.g., Graham 1985; Tobin 1984; Schoenbrod 1983; Ackerman and Hassler 1981; LaPierre 1977; and Silver 1976–1977).

Despite the fact that environmental laws have not fully achieved their ambitious statutory goals (perhaps few laws do), there is little sign that either the laws or their basic structures will be abandoned, at least in the near future. There has been continuing deadlock in Congress for several years over the reauthorization of several major environmental laws. For example, the Clean Air Act has been finally reauthorized after over a decade of effort; in 1987, the Federal Insecticide, Fungicide and Rodenticide Act, after a more than a decade, was reauthorized; and in the same year, the Clean Water Act was reauthorized after about five years of effort. Thus, the basic statutory structure combined with entrenched interests pose fundamental constraints for further policy making efforts.

Even though there is wide recognition in Congress that: (1) certain scientific principles upon which the laws rest are not unambiguously valid now, if they ever were; and (2) there is a generally recognized need for more adequate and explicit language in the statutes to specify how health and environmental risks and technological and economic costs should be balanced in the standard-setting process (e.g., Majone 1984; Schoenbrod 1983); it is not likely that significant clarification of the laws on these basic issues will occur. The political interests in these policy arenas are well-entrenched almost twenty years after the laws were passed.

This has implications for how Congress addresses the other key challenge facing it: the issue of scientific uncertainty. For most environmental policy issues, scientific and technical information is likely to be part of the "weaponry of political conflict," rather than used to help resolve policy disputes. Although the various interests may disagree about the validity of the current laws, they support maintaining the status quo rather than risk changes they may support less.

The dynamics of the early 1970s led Congress to adopt statutes, such as the Clean Air Act, which seem to "avoid compromise by stating an absolute duty to achieve a set of goals—the protection of health, welfare, and natural air quality. But Congress did not decide how the burden of achieving them would be allocated" (Ackerman and Hassler 1981). The administrative process for implementing the law has become extremely lengthy and complex, yet most reform efforts focus on making the laws more "efficient" or "workable," rather than on the basic statutory structure itself, the ultimate source of most implementation problems. It is likely that debates over deadlines for achieving the current statutory goals and relatively minor refinements will continue to dominate the reauthorization efforts. Shep Melnick (1984; 1983) has pointed out that such debates have benefits for all the actors involved in the policy process.

For example, the symbolic quality of the current statutory framework of strict deadlines (and broad delegations) is valuable for Congress: the belief is that they encourage agency action which makes them supportable by environmentalists, and the prospect that they will not be met means that industry interests find them basically innocuous. Yet, reforms could be contemplated which would temper the broad, symbolic goals of the laws and allow some explicit balancing of environmental and other competing public goals, and reforms which would simplify some unnecessarily restrictive statutory provisions and, where needed, add flexibility for agency action.[2]

The persistence of environmental concern in the nation, despite the economic downturns of the mid-1970s and the all-out attack of the early Reagan administration, is strong testimony to the staying power of the environmental laws. There are, however, noticeable differences in the basic design of some of the earlier environmental laws and those written in the 1980s. This is due in part to the realization by the late 1970s that the expectation for science and technology to meet the broadly stated goals of the environmental laws was unfulfilled. It also reflects the most recent form of federalism which defines a limited federal role and greater state responsibility for developing and implementing policies (Rabe 1988). The two cases reviewed in this paper will illustrate these differences.

Use of Science

Throughout the century, the federal government's goals and concomitantly its responsibilities have mushroomed. Although there are many reasons for this expansion, it is clear that science and technology helped fuel it. It was, for example, our increased ability to identify and study microorganisms by the early 1900s that sensitized us to the health hazards of polluted water and led to the early water pollution laws. Today, our increased ability to detect minute quantities of toxins has spurred further regulation. One of the most striking features of the current pollution laws is their heavy reliance on technological solutions for pollution abatement.

Congress has found it increasingly convenient to use the language of science and to rely on it and new-found technologies for solution to a whole range of public problems. As Harvey Brooks observed this now means that "there is constant pressure to convert political questions into technical questions, so that they can be referred to experts without actually confronting the value differences that frequently are the real origins of conflict" (Brooks 1984). The intention is to use the apparent objectivity of science to suggest that technical rather than political choice determines the regulatory policies (Dickinson 1984).

Environmental laws of the 1970s and 1980s epitomize this tendency with their emphasis on finding the best technical solutions to pollution. In 1970, Congress basically legislated that there should be "clean air" by 1977. The job of choosing controls and rules of conduct was delegated to EPA (Schoenbrod 1983; Tobin 1984). Congress, in many instances, did not provide guidance for the kinds of trade-offs between economic and environmental concerns, or for the balancing of risks based on uncertain evidence, which are fundamental elements of environmental policy making. At a minimum, this has left EPA officials struggling to interpret an overwhelming amount of scientific, technical, and economic information with inadequate resources, despite their enormous responsibilities.

There has always been a close association between environmentalism, science, and public policy in that "environmentalists have sought to generate public demand for policies that are suggested by scientific findings" (Paehlke 1989). Rachel Carson's publication of *Silent Spring* in 1962 established one of the first critical links between scientific knowledge of an environmental problem and needed political action. Carson's work, building on the work of Leopold and carried on by Barry Commoner and other contemporary scientists/advocates, was intended to "persuade as well as inform" (Graham, as quoted in Paehlke 1989, 28).

Robert Paehlke (1989, 36) in his recent work on environmentalism persuasively argues that while the "symptoms of environmental problems may be measured biologically . . . the disease itself lies in our socioeconomic organizations, *and the solutions are ultimately political"* (emphasis added). Ultimately, a combination of societal dispositions and scientific constraints appear to have unwittingly conspired to prevent timely political action based on available scientific information on the effects of various toxic substances.

In any case, the environmental sciences (i.e., ecology, toxicology, and epidemiology) today have a fundamental role in environmental policy making. With this, according to Paehlke, a fundamental transformation in the purpose of science in society is evolving. After noting that science and technology for centuries have served as the means for accelerating economic advance, he observes that science "can continue to advance productive efficiency, but that efficiency must [now] be seen in organic and ecological rather than merely mechanistic terms. [Specifically, e]nvironmental science assumes that every new technology introduces undesirable and commonly unanticipated impacts." For this reason, the benefits and costs of actions must be weighed. Consequently, "economics is at the heart of decisions regarding the application and evolution of science, and it is central also to the gulf within science" between the traditional and new purposes of Science (Paehlke 1989, 118, 119).

It is an open question how well scientists will assume the roles for the new type of science. It will necessitate scientists to "not only investigate but inform, not only advise but dissent when appropriate, and, most importantly, provide firm, prudent leadership in the formulation of public health policies" and other policy issues as well (Wagoner, as quoted in Paehlke 1989, 34). It also remains to be seen how capable or motivated policy makers are to better ground policy decisions on *available* scientific and technical information.

The creation of the Office of Technology Assessment (OTA) by Congress in 1972 and the re-direction of the General Accounting Office (GAO) and the Congressional Research Service (CRS) to oversight and investigative studies represent attempts by Congress to use its agencies as forums of policy-oriented scientific and technical assessments of environmental and other policy issues. OTA undertakes assessments of policy issues at the request of congressional committees and with the approval of its congressional board, the Technology Assessment Board. The assessments are usually timed with the reauthorization of major legislation or in an area for which Congress is considering new legislation. Although numerous sources of scientific and technical information exist, a primary purpose of OTA is to be Congress' own independent source of such information. Although OTA's assessments are usually seen as key components of policy debates, as with any information presented in the policy process, how it is used is largely beyond the control and prediction of its source.

CRS, which is part of the Library of Congress, is another congressional agency. Usually, it responds with brief reports to requests from members of Congress or from congressional committees. It also may facilitate the development of hearings, speech writing, etc. GAO since the 1970s has been encouraged by Congress to take on a "watchdog" approach (i.e., to initiate its own investigations of agency implementation efforts). Another congressional support agency is the Congressional Budget Office. Together these agencies are intended to provide Congress with various kinds of technical information assessments needed to facilitate congressional policy making.

Clearly, the interpretation of scientific information and the uncertainty often associated with scientific understanding of environmental issues are important challenges for congressional policy making today. Congressional agencies provided important studies in both cases of environmental policies which are reviewed in this paper. Yet the two cases—at least in the initial phase of congressional action—provide contrasting approaches to the use of scientific and technical information and the reliance on technology-based solutions by Congress.

A Case of "Seeing Clearly"

The major statute addressing air pollution is long: the Clean Air Act of 1977 accounts for 118 pages of the *United States Code* and for four volumes of the *Code of Federal Regulations*. As time passes and the statute continues to grow, the reauthorization process becomes increasingly long and laborious, as evidenced by the 1990 Amendments to the Act. The administrative process, too, has become lengthy and involved. What started as a thirteen step process in 1971 to set national ambient air quality standards (NAAQSs), by the mid-1980s grew to a 53 procedural step process taking years to complete with no definite policy actions resulting (Berry 1984).

A major feature of the Clean Air Act of 1970, basically unchanged in subsequent legislation, is Section 109 which requires EPA to establish uniform NAAQS for certain pollutants and directs states to implement them, with EPA's approval and assistance. There are five basic stages in NAAQS system:

- Goal-setting (i.e., targets for abatement);
- Criteria development (i.e., synthesizing scientific and technical data of the pollutant to identify human health and environmental effects);
- Air quality standard-setting (i.e., maximum allowable levels of pollutant in ambient air based on the criteria document; set for the protection of both human health (primary standards) and the environment, public welfare, etc. (secondary standards);

- Emission standard-setting (i.e., delineating the permissible levels for the pollutant emission source); and
- Enforcement (i.e., ensuring compliance with sanctions, etc.) (Regens and Rycroft 1988).

Seven NAAQs have been set to date for pollutants linked to such air quality problems as acid rain and urban ozone pollution (i.e., smog).[3] The scientific defensibility of several of the standards is questionable, but they are difficult to set—and to change. Each stage of the standard-setting process is wrought with conflicts.

Much of the current debate centers on which criteria are most appropriate for setting the air quality standards: economic efficiency, environmental effectiveness, or social equity (Regens and Rycroft 1988; OTA 1989). In addition, tensions between the federal and state governments, and in turn between state and local governments, have hampered implementation. The statute establishes the broad, societal goal of achieving air quality necessary to protect the public health and "allowing an adequate margin of safety" with little guidance to EPA on how to weigh the costs, risks, and benefits associated with any regulatory action. Further, the deadlines for achieving such goals (originally, set for 1975) were extremely unrealistic. They were rather arbitrarily set and underestimated the scientific and technical difficulties in performing the tasks necessary to set standards and control sources.

Precisely because of the emphasis on science and technical solutions in most environmental statutes, the EPA relies heavily on the judgments of its science advisory committees to provide legitimacy for its implementation strategies and actions. One consequence is that at times these committees have assumed considerable authority within the Agency. The Clean Air Scientific Advisory Committee (CASAC) was created by the Clean Air Act and subsequently became a part of the EPA's Science Advisory Board (SAB). Although it is beyond the scope of this paper to examine the advisory process within the EPA, it is important to note that at times these committees, and particularly CASAC, have had a determining effect on the standard-setting process.

Advisory committees have been credited with causing delays in the standard-setting process and taking the initiative to assume greater reviewing authority within the agency (GAO 1984; 1983). In fact, CASAC members and EPA officials were called at least once before a congressional oversight committee hearing to explain their apparent "more than just consultation role" of CASAC in the EPA (Hearing 1983). The apparent heavy reliance by the EPA on its science advisors is hardly surprising given the generally open-ended nature of the statute and the complexity of the issue.[4]

In 1989, to facilitate Congress with the ongoing reauthorization of the Clean Air Act, the Office of Technology Assessment released a technology assessment on the urban ozone issue which helps distinguish questions of scientific certainty, technical capabilities, and political choices for Congress (OTA 1989). The Congressional Research Service also worked with the relevant congressional committees throughout the decade-long reauthorization process for the Clean Air Act.

As these studies explain, urban (not to be confused with stratospheric) ozone is considered a major source of air pollution in many areas of the country, yet the NAAQS set by the EPA to control ozone is not being met in cities comprising approximately half of the country's population (OTA 1989). Ozone results when its precursors, volatile organic compounds (VOCs) and nitrogen oxides, combine in sunlight. The result is a haze, often referred to as smog. VOCs is a category of hundreds of pollutants, including human sources (e.g., vehicle exhaust, evaporation of solvents and gasoline, chemical manufacturing, and petroleum refining) and sometimes (particularly in the summer) natural vegetation. Nitrogen oxides primarily emanate from fossil fuel combustion (e.g., vehicles, and utility and industrial boilers). To date, ozone control efforts have focused on reducing local VOCs because the relevant technologies were thought to be more readily available and cheaper (OTA 1989). Yet, given the meager progress to date in achieving the ozone standard, additional measures are needed.

OTA points out that while the current national standard (0.12 ppm) rests on solid evidence from health effects of short-term exposure above that level, "experts are unsure" whether the standard adequately protects people who are exposed for long times or at high exercise levels. The "connection between irreversible lung damage and repeated exposure to summertime ozone levels remains inconclusive," yet, "clear evidence" exists that ozone damages "economically, ecologically, and aesthetically important plants." Further, the questions of how serious the damages and risks are, how much it will cost to avoid them, and how the costs will compare to the benefits are all ones that scientists cannot answer with certainty. As OTA rightly notes, "Deciding how to act in the absence of full information falls to Congress and the Nation" (OTA 1989).

The technology assessment then goes on to discuss various types of options Congress has "to balance the urgency of the ozone problem against the difficulty of the solution" and make the necessary policy decisions in the face of scientific uncertainty. As part of this discussion, the costs and impact on reductions of different sources are compared and assessed. Given that certain options could entail significant lifestyle changes (e.g., restrictions on the use of vehicles or requiring the use of electric vehicles), it may be political resistance renders some options nonviable.

The assessment also reviews in detail the health effects of ozone, effects of ozone on crops and forests, as well as discussing the various opportunities and options for controlling ozone (i.e., through controlling volatile organic compound and nitrogen oxide emissions) (OTA 1989).

Congress addressed these and other issues, which already had been debated for over a decade in Congress, and finally reauthorized the Clean Air Act in 1990. How quickly implementation, let alone when actual impacts on ozone levels, will occur is highly uncertain. Equally uncertain is how adequately Congress utilized the greater availability of information today on the nature of ambient pollution issues and the trade-offs their abatement requires.

The main reasons given for the meager implementation results to date for attaining the ozone standard include: (1) incomplete or inaccurate emission inventories (used to estimate current emissions of ozone precursors and to project future emissions in the absence of additional controls); (2) underestimates of control measures required (based on modelling) to meet the standard; (3) underestimating emissions growth and poor control over it; (4) difficulty of states in issuing stationary source regulations; (5) difficulty in meeting unrealistic statutory deadlines; and (6) a lack of political leadership and will to act decisively (OTA 1989; NCAQ 1981). The basic statutory structure effecting these factors has not been changed significantly by Congress. Thus, the scientific and technical difficulties which exist and foil much of the implementation to date do not obviate how the statute itself is a key source of the present problems and may remain key to the prospects for improving implementation efforts.

A "Vital Signs" Case?

The beach wash-ups of medical wastes in recent summer seasons drew attention to the confusing, inconsistent and inadequate regulation of these wastes.[5] At the federal, state, and local levels of government a number of responses have been made to better address the management of medical wastes. Prior to 1988, the federal government had not regulated medical waste management (OTA 1988; Church et al. 1989). There is general consensus that proper handling, treatment, and disposal of medical wastes will present minimal human health and environmental risks. Yet, obviously, incidents of careless of illegal disposal may pose health risks and aesthetic problems and certainly help create public apprehension over current medical waste management practices. In addition, potentially significant health and environmental risks may result from poorly operated and maintained existing medical waste incinerators with few or no air emission controls (OTA 1988; 1990).

A background paper, *Issues in Medical Waste Management*, was issued by OTA in October 1988 as part of a larger assessment of municipal solid waste (MSW) issued in October 1989. The OTA study addressed the adequacy of current medical waste disposal practices, the potential risks from such practices, and the need for further federal requirements for the handling, treatment, storage, and disposal of medical wastes. It was referred to by the congressional staffers as they drafted the Medical Waste Tracking Act (Church et al., 1989).[6] OTA also testified before Congress and consulted with Congressional staff on medical waste issues.

As the OTA (1988) study of medical waste issues noted, "Currently, no federal regulations exist that *comprehensively* address the handling, transportation, treatment, and disposal of medical waste." Further, the federal guidelines that do exist are inconsistent in their definition of medical/ infectious waste and in their suggested management options. The EPA, Centers for Disease Control (CDC), and other federal agencies have issued different guidelines for infectious and medical waste management. This, of course, means that states have largely been left on their own to devise medical waste management programs and thereby important variation frequently exists between state requirements, as well as between local requirements and those of a state.

The Medical Waste Tracking Act (MWTA) of 1988 is an attempt by Congress to address the problems of beach wash-ups and illegal disposal of medical wastes.[7] A more comprehensive approach to medical waste management could be established if the issue of medical wastes remains part of the Research Conversation and Recovery Act (RCRA) reauthorization effort, or if the demonstration program of the MWTA is expanded and extended in the future.

The MWTA defines medical waste as "any solid waste which is generated in the diagnosis, treatment, or immunization of human beings or animals, in research pertaining thereto, or in the production or testing of biologicals."[8] Although the medical waste stream is extremely heterogeneous, the focus has been on the portion of the waste stream termed "infectious," and how these "red bag" wastes are classified (e.g., solid, hazardous, or special) and regulated. Most estimates are that 10–15% of medical waste are infectious, although the range can be for up to 80% of medical waste to be classified as infectious depending on the generator's definition. Determining which portion of medical waste is infectious remains at the heart of definitional issues and how infectious waste is defined can greatly affect the cost of waste management, and ultimately the choice of disposal options (OTA 1988).

Various federal entities have differing definitions of infectious wastes (OTA 1988). However, the definitions of EPA and the MWTA, CDC and the Occupational Safety and Health Administration (OSHA) agree of four *infectious* waste categories: pathological waste, blood and blood products, contaminated sharps, and microbiological wastes. The EPA as directed by the MWTA issued regulations in June 1989 for a two-year demonstration medical waste tracking program, which included categories of "medical waste" as:

- cultures and stocks of infectious agents;
- human blood and blood products;
- human pathological wastes (including those from surgery and autopsy);
- contaminated animal wastes;
- wastes from patient isolated with highly communicable diseases; and
- all used sharps and certain unused sharps.

Thus, although no consensus exists yet on the categories of wastes designated as infectious, it is possible that the EPA definition as codified as part of the MWTA might become more widely used.

MWTA establishes a demonstration tracking system (Sections 11001–11003) and directs the EPA and the Agency for Toxic Substances and Disease Registry to undertake studies of certain medical waste management issues (Sections 11008 and 11009). The intent is to develop a basis for determining, after the completion of the demonstration program, whether and in what ways the federal government should regulate medical wastes. MWTA specifically applies to Connecticut, New Jersey, New York and the Great Lakes states (Section 11001). Any of the Great Lakes States could opt out of the demonstration program and any other state could opt in; Connecticut, New Jersey, and New York could petition out if their state has a program at least as stringent as that of the federal government. Although none of the three states petitioned out, each of the Great Lake states opted out. This means that the MWTA applies only to New York, New Jersey, Connecticut; and additionally, Rhode Island and Puerto Rico, which voluntarily entered the program.

The tracking system for medical wastes devised by the EPA involves record keeping requirements for facilities which generate over 50 pounds a month of medical waste and requires the use of a four-part form for any offsite shipment of medical wastes. If a generator does not receive a copy of the manifest form from the final destination facility, the EPA and the state must be notified. Generators of medical waste that produce less than 50 pounds are subject to the same handling requirements, except that they must maintain a log (i.e., a reporting requirement) instead of using the tracking form.

Congress is well aware that a number of issues regarding medical waste management remain unresolved. These include issues regarding proper treatment, appropriate technologies and controls, and feasible disposal for home-generated medical wastes. It may be that the studies required by the MWTA will assist in resolving some of these issues. In addition, the importance of the federal demonstration tracking program for medical wastes to abating medical waste problems is not clear. The contribution of a tracking system to the improved management of medical waste will need to be evaluated independently which is expected to be done based on reports to Congress by the EPA upon the completion of the demonstration program.

The EPA has authority under RCRA to regulate the handling, storage, treatment, transportation, and disposal of medical wastes.[9] It remains an open question how extensively the federal government will be involved in devising such an approach. As noted, Congress is mindful of the fact that a number of issues remain which require further examination in order to best support any regulatory activity.[10] After the completion of the demonstration project, it is scheduled to reconsider the need for further federal regulation of medical wastes. In the past, Congress has been highly respectful of state and local authority for managing the MSW and medical waste streams. Although the national implications of managing these wastes and the variations in state and local regulations for them are becoming increasingly important, the federal government is responding cautiously as it attempts to determine its appropriate policy role.

A follow-on study by OTA was requested by several congressional committees shortly after the passage of MWTA. This resulted in an OTA special report to Congress in 1990 (OTA 1990). The focus of the study is to assess a number of incineration and other medical waste treatment technologies issues and to examine medical waste reduction and recycling opportunities. The General Accounting Agency (GAO), another congressional support agency, completed two short studies requested by congressional committees (GAO 1989; 1990). One study focused on the medical waste management by federal facilities. The other study provided a preliminary analysis of EPA's implementation of the MWTA (i.e., establishing one demonstration tracking program) and evaluated the regulatory programs of several states for managing medical wastes.

The intention of the congressional committees is that information gained from the reports by Congress' own agencies, along with that assembled by the EPA and the Agency for Toxic Substances and Disease Registry (ATSDR) and in addition to information gained from the demonstration project, from states and local governments, and from private sources, will be used by Congress to evaluate the need for a more permanent program and the appropriate federal role in the management of medical wastes.[11]

While the Senate Committee on Environment and Public Works and the House Committee on Energy and Commerce (and their relevant subcommittees) were immersed in the reauthorization of the Clean Air Act, they also began contemplating what form the RCRA reauthorization effort should take as the next major item on the environmental agenda for Congress. Given that the federal role has been limited in the past with respect to MSW, and particularly medical wastes, policy interests are not as entrenched at the national level for these issues as they are for air pollution issues. It appears in light of the past experience with ambitious legislative responses to newly discovered environmental problems, which led primarily to a slow and cumbersome implementation process, as well as the fact the current state of federalism is biased against the initiation or assumption of large federal programs, that Congress is proceeding cautiously with respect to the medical waste problem. Whether this translates into the use of scientific and technical information more for clarification than as "weaponry," remains to be seen. The contrast between the congressional response to this environmental issue and that of air quality nearly two decades ago is striking. For Congress, both issues have prominence given their frequent coverage by the media and the high level of public concern about them. For both issues the level of scientific certainty about the nature of the problem and risks associated with it were fairly high. The congressional response for each, however, was quite different.

Conclusion

It appears from these cases that the relevance of scientific and technical factors to environmental policy making decisions today may vary depending on the novelty of the problem for Congress and the degree of entrenchment of interests concerned with any policy solutions to it. Under such conditions, it is for newer issues that Congress may be most likely to better base legislation on available scientific knowledge and technical capabilities than it has with older issues. This may allow for greater flexibility to be structured into the law, which will better allow for adjustments of strategies as new information becomes available.

It is well to remember that "political disputes require political resolution" (Roberts et al. 1984). Roberts et al. have distinguished several propositions regarding the likelihood of resolving scientific disputes which affect policy making.[12] Ultimately, the origins of impasse in most environmental policy areas are basically political, not technical in nature. The technical complexity or scientific uncertainty surrounding a particular issue or aspects of it may increase the difficulty of policy making, but political willingness and other motivations of the key participants will be more determinative of the policy itself. It is likely then that the better established the issue network (using Hugh Heclo's term), i.e., those government and non-government entities most attentive and interested in a particular policy's development, the more extensively scientific and technical information is used for political purposes.

That is, science is most likely to be part of the "weaponry of political conflict" (Roberts et al. 1984) in an entrenched policy arena. In these cases, I agree that "too many of the participants have good reasons *not* to distinguish scientific evidence from policy preferences, *not* to analyze carefully the various sources of technical disagreement, and *not* to accept responsibility for some decisions or judgments" (Roberts et al. 1984). Yet, in recent policy debates, particularly for new issues before Congress, it may be that the legislation reflects greater appreciation of what science can and cannot contribute to resolving policy disputes.

In the case of NAAQS standards, the EPA created by design or default an elaborate set of procedures to contend with the inevitable challenges to their proposals.[13] The Clean Air Act of 1970 granted the EPA broad authority to establish stringent standards, but avoided determining what weight the Agency would have to give to certain key criteria, such as the economic cost of controlling the emissions of a pollutant. An elaborate science advisory role evolved, but basic scientific uncertainty remains, as well as uncertainty over a viable policy outcome. The EPA has little flexibility under the statute to devise, for example, an ozone standard which is likely to withstand challenge given the extensive opportunities for citizen participation in the law. With so much attention on the outcome (or lack thereof), little has been focussed on the process used to attempt a resolution to the policy issue.

The elaborate process which has evolved was indirectly encouraged by the nature of the statute. The process has encouraged use within the Agency of scientists, science, and technical information as buffers in the policy making process. A discussion of the role of science advisory panels within the EPA is beyond the scope of this paper. Yet, it is important to note that the problems of distinguishing what are political and what are technical questions, and what science can and cannot contribute to the resolution of policy disputes are addlepated within this process.

In the case of medical waste management, Congress has taken a much more cautious route. It passed legislation in 1988 which established a two year demonstration program and a research agenda for EPA and ATSDR. In addition, it requested further study by the OTA of the issue and studies by another congressional support agency, the General Accounting Office (e.g., GAO 1989) as well. Based on the information gathered by all of these efforts, Congress will re-evaluate the issue. The attempt is to clearly define the policy goal of ensuring safe management of medical wastes and structure a process to gather sufficient information to evaluate the nature and extent of risks posed by these wastes in order to later identify desirable and viable policy options.

It is evident from past experiences in grappling with the scientific uncertainty and technical complexity inherent in most environmental policy making that while scientists need to be explicit about how policy concerns impact scientific and technical disagreements, policy makers also need to be explicit about which *policy* choices exist amongst competing goals and despite scientific and technical uncertainties and disagreements (Roberts et al. 1984; Marcus 1988). In all likelihood decisions will be necessary despite uncertain knowledge and a lack of agreement over prudent courses of action. Responsible policy making requires that policy decisions be made in this context by Congress in order to establish the rules of conduct for agency action (Schoenbrod 1983; Lowi 1979). Technology assessment, in this context, can be an appropriate form of policy analysis to facilitate congressional policy making. In both the cases discussed, studies of various scientific and technical aspects of the policy issues and interpretations of them by congressional support agencies assisted Congress and policy making efforts in the 1980s. This is in striking contrast to the situation of congressional policy making in the 1970s. Scientific information and studies available to Congress during the 1970s were typically "not intended to meet the test of scientific judiciousness but rather to satisfy specific legal and administrative requirements" (Landy et al. 1990).[14] That is, most of these studies were in the mode of "advocacy science." The studies provided by congressional agencies attempt to present information to congressional committees from as unbiased and nonadvocacy position as possible. This suggests an important role for these agencies in elevating the ability of Congress to distinguish scientific and technical issues from political ones and better address the former as such.

Even in the policy areas for which political interests are well entrenched, it may be that greater insistence by all participants that science and technical issues and political issues be clearly distinguished in their importance to forging a policy resolution will result in more responsible policy making. Policy making in a democratic society has never required consensus on all matters, but it does require clarifying disagreements and acting in the face of risks, controversy, and uncertainty to be responsible.

Notes

1. Over ten major environmental statutes were enacted during the 1970s, over half of these were passed in the first half of the "environmental decade," as it came to be called.

2. See generally, for example, Schoenbrod's (1983) discussion of "goals statutes" and "rules statutes" and his arguments supporting a movement toward drafting the latter.

3. These are: particulate matter, sulfur dioxide, carbon monoxide (standard eliminated by EPA in 1983), nitrogen dioxide, photochemical oxidants (i.d., ozone), hydrocarbons, and lead.

4. Central to the relationship between the EPA and the SAB/CASAC is the closure statement process. As a CASAC document explains, "Closure represents a sense of the committee determination upon the scientific adequacy of a criteria document for regulatory purposes at a specific point in time, based upon information currently available." Frequently, a closure letter for a pollutant will not be written until the document under question has incorporated SAB/CASAC's comments. Although EPA offices are not legally bound by the closure process, they have tended to make the necessary changes and return to SAB/CASAC, often several times, to obtain closure (GAO 1983; Hearing 1983). According to CASAC, closure "provides a strong incentive for cooperation between the Agency and the [advisory] committee" (CASAC 1981).

5. Medical wastes are all infectious, hazardous (including low-level radioactive) and any other wastes (as defined by law) generated from all types of healthcare institutions (includes hospitals, clinics, doctor, dental and veterinary offices, hospices, nursing homes, funeral homes, and research laboratories). It is still difficult to pinpoint the amount of medical waste being generated nationally. Roughly, the range is between 16–23 lbs./bed/day for *hospital* waste which is about one to two percent of the total amount of MSW generated. It is not clear what amount of medical waste is generated from other sources (e.g., home healthcare, etc.) (OTA 1988).

6. Most of the initiative for the MWTA originated in the House, although the Senate Environment and Public Works Committee (with several members from coastal states which experienced beach washings of medical wastes) was active on the issue and several ideas from its bills were introduced into the MWTA and other related legislation.

7. The Medical Waste Tracking Act is amended as new Subtitle J to the Solid Waste Disposal Act and the Resource Conservation and Recovery Act (42 U.S.C. Sections 6901 et seq.). Congress also amended the Ocean Dumping Act in 1988 to increase the penalties for illegal disposal of medical wastes by public vessels. The Ocean Dumping Act is formally known as the Marine Protection, Research, and Sanctuaries Act of 1972 (33 U.S.C. Sections 1401 et seq.).

8. Section 1004(40) of RCRA, as amended by MWTA, notes further that medical waste "does not include any hazardous waste identified or listed under Subtitle C or any household waste as defined in regulations under Subtitle C." Any solid waste which is mixed with a regulated (i.e., listed) medical waste (see below) is also a regulated waste under the MWTZ. It should also be noted that domestic sewage is not included in the RCRA definition of solid waste (Section 1004(27)).

9. Before the passage of the MWTA, the EPA's activity was mostly limited to distribution of its guidance document for the management of infectious wastes. Other medical wastes were considered to be like any other solid waste and subject to relevant RCRA Subtitle D regulations.

10. For example, three key challenges which need to be addressed are: 1) improving the definition and identification of infectious and other medical wastes to facilitate more consistent and adequate handling and treatments of wastes; 2) better addressing the diversity of generators (e.g., home healthcare, small doctor offices, clinics, etc.) to minimize contradictory procedures and inequities posed by them; and 3) improving the segregation of wastes for their proper treatment.

11. It is interesting to note that one of the most active committees on the medical waste issue during this time was an *oversight subcommittee* of the House of Representatives Committee on Small Business. Representative Ron Wyden (D Ore), chair of the Subcommittee on Regulation, Business Opportunities and Energy, (and his staff) had an active interest in the medical waste issue—a high visibility issue in 1988. The Subcommittee had recently completed investigating medical testing and it appeared to them that the medical waste issue meshed logically as a follow-on to that. Although Wyden is a member of the House Energy and Commerce Committee, the committee with primary *legislative* authority over the medical waste issue, he is not a chairman of a subcommittee of that Committee. Wyden's Subcommittee was a requestor of the second OTA report and requested the GAO reports. In addition, it held more hearings on the issue than the appropriate subcommittee of the Energy and Commerce Committee.

Indeed, the Energy and Commerce Subcommittee on Transportation of Hazardous Materials attempted to keep a low profile on the issue. Initially, the Committee, chaired by Tom Luken (D-Ohio) cautiously explored the topic as an aside to the municipal solid waste (MSW) issue and held hearings in the summer of 1987 (Hearings, 1987, OTA, 1987). In discussions with OTA staff working on the comprehensive assessment of MSW, in anticipation of the upcoming reauthorization of RCRA, the subcommittee requested that the medical waste portion of the study be published separately to facilitate development of what became the MWTA. Both Subcommittees and the House Science, Space and Technology Committee (another requestor of the MSW study by OTA) requested the second OTA report on medical wastes.

12. The four basic hypotheses posed are: 1) "Political disputes require a political resolution;" 2) "Scientific conflict resolution is easiest when there is the least scientific conflict;" 3) "Direct bargaining or mediation efforts are worthwhile only when it is clear who should participate, and when participants are willing to reach, and able to maintain, an agreement;" and 4) "For powerful administrative agencies, the consequences of any disputeresolving or dispute-clarifying process will depend more on the details of how the process is implemented than on which class of participants it involves" (Roberts et al., 1984). Although in this paper I will not attempt to "test" any of the hypotheses of Roberts et al., recognizing them more as insights than testable propositions, the cases reviewed do appear to support many of the same conclusions regarding the role of science as technical information.

13. As Harris and Milkis (1989) note, the Clean Air Act of 1970 and other environmental laws passed in the early 1970s changed regulatory politics in two ways: 1) procedurally, they altered the issue networks and institutional bases of regulatory politics by promoting the ideal of participatory democracy; and 2) substantively, they were aimed at achieving environmental objectives. The "drive

toward participatory democracy" manifested itself in this policy area in the public hearing requirements and citizen suit provisions.

14. Another example of this is associated with the passage of the Comprehensive Environmental Responses, Compensation, and Liability Act (CERCLA) of 1980, commonly referred to as Superfund. The first OTA report on Superfund was published in 1985 (OTA 1985). It provided a much broader, more independent scientific and technical basis for congressional reauthorization of that law in 1986, the Superfund Amendments and Reauthorization Act (SARA).

Bibliography

Ackerman, Bruce A. and William T. Hassler. 1981. *Clean Coal/Dirty Air or How the Clean Air Act Became a Multibillion-dollar Bail-Out for High-Sulfur Coal Producers and What Should Be Done About It*. New Haven: Yale University Press. Beam, David. 1981. "From Law to Rule: Exploring the Maze of Intergovernmental Regulation." *Intergovernmental Perspectives* 9:7–22.

Berry, Michael. 1984 (May). "A Method for Examining Policy Implementation: A Study of Decisionmaking for the National Ambient Air Quality Standards, 1964–1984." Washington, D.C.: U.S. Environmental Protection Agency.

Bosso, Christopher J. 1987. *Pesticides &0 Politics: Life Cycle of a Public Issue*. Pittsburgh: University of Pittsburgh Press.

Brooks, Harvey. 1984. "The Resolution of Technically Intensive Public Policy Disputes." *Science, Technology, and Human Values* 9(1):39–50.

Church, Thomas W., Phillip J. Cooper, and Robert Nakamura. 1989. "The Political and Regulatory Environment of Medical Waste: Formulation and Implementation of the Medical Waste Tracking Act." Report to the Medical Waste Policy Committee. Albany, New York: The Nelson A. Rockefeller Institute of Government, State University of New York.

Dickinson, David. 1984. *The New Politics of Science*. New York: Pantheon Books.

Graham, John. 1985. "The Failure of Agency-Forcing: The Regulation of Airborne Carcinogens Under Section 112 of the Clean Air Act." *Duke Law Journal* 1985:100–150.

Harris, Richard A. and Sidney M. Milkis. 1989. *The Politics of Regulatory Change: A Tale of Two Agencies*. New York: Oxford University Press.

Heclo, Hugh. 1978. "Issue Networks and the Executive Establishment." *The New American Political System* edited by Anthony King. Washington, D.C.: American Enterprise Institute.

Jones, Charles. 1975. *Clean Air: The Policies and Politics of Pollution Control*. Pittsburgh: University of Pittsburgh Press. Landy, Mark, Marc J. Roberts and Stephen R. Thomas. 1990. *The Environmental Protection Agency: Asking the Wrong Questions*. New York: Oxford University Press.

LaPierre, Bruce. 1977. "Technology-Forcing and Federal Environmental Protection Statutes." *Iowa Law Review* 62(3):771–338.

Library of Congress. Congressional Research Service. 1989. *Health Benefits of Air Pollution Control: A Discussion*. Washington, D.C.: Library of Congress.

———. 1990. *Potential Benefits of Enacting Clean Air Amendments.* Washington, D.C.: Library of Congress.

Lowi, Theodore J. 1979. *The End of Liberalism: The Second Republic of the United States.* 2nd ed. New York: W.W. Norton and Company.

Majone, Giandomenico. 1984. "Science and Trans-Science in Standard Setting." *Science, Technology, & Human Values* 9(1):15–22.

Marcus, Alfred. 1988. "Risk, Uncertainty, and Scientific Judgment." *Minerva* XXXVI:139–152.

Melnick, R. Shep. 1983. *Regulation and the Courts: The Case of the Clean Air Act.* Washington, D.C.: The Brookings Institution.

———. 1984. "Pollution Deadlines and the Coalition for Failure." *The Public Interest* 75:123–134.

Mitchell, Robert. 1984. "Public Opinion and Environmental Politics in the 1970s and 1980s." In *Environmental Policy in the 1980s*, edited by Norman Vig and Michael Kraft. Washington, D.C.: CQ Press.

Moore, John L. 1990. "Policy Analysis in a Legislative Setting: Some Perspectives from the Environmental Arena." Paper prepared for the Twelfth Annual Research Conference of the Association of Policy Analysis and Management, October 18–20, 1990, San Francisco, California. National Commission on Air Quality. 1981 (March). *To Breathe Clean Air.* Report of the National Commission on Air Quality. Washington, D.C.: National Commission on Air Quality.

Paehlke, Robert C. 1989. *Environmentalism and the Future of Progressive Politics.* New Haven: Yale University Press.

Rabe, Barry G. 1988 (Fall). "Toward the New Federalism." *Forum for Applied Research and Public Policy* (Fall):5–15.

Regens, James L. and Robert W. Rycroft. 1988. *The Acid Rain Controversy.* Pittsburgh: University of Pittsburgh Press.

Roberts, Marc J., Stephen R. Thomas, and Michael J. Dowling. 1984 (Winter). "Mapping Scientific Disputes That Affect Public Policymaking." *Science, Technology, & Human Values* 9(1):112–122.

Schoenbrod, David. 1983. "Goals Statutes or Rules Statutes: The Case of the Clean Air Act." *UCLA Law Review* 30(4):740–828.

Silver, Jonathan. 1976–1977. "Problems in Attempting to Translate Statutory Standards into Emission Limitations Under Air and Water Pollution Controls Legislation." *Villanova Law Review* 22:1122–1170.

Stone, Alan. 1980. *Regulation and Its Alternatives.* Washington, D.C.: CQ Press.

Tobin, Richard. 1979. *The Social Gamble: Determining Acceptable Levels of Air Quality.* Lexington, Massachusetts: D.C. Heath.

———. 1984. "Revising the Clean Air Act: Legislative Failure and Administrative Success." In *Environmental Policy in the 1980s: Reagan's New Agenda*, edited by Norman Vig and Michael Kraft. Washington, D.C.: CQ Press.

U.S. Congress. Office of Technology Assessment. 1985 (April). *Superfund Strategy.* Washington, D.C.: GPO. ———. 1988 (October). *Issues in Medical Waste Management: Background Paper.* Washington, D.C.: GPO.

———. 1989 (July). *Catching Our Breath: Next Steps for Reducing Urban Ozone.* Washington, D.C.: GPO.

———. 1989 (October). *Facing America's Trash: What Next for Municipal Solid Waste?* Washington, D.C.: GPO.

———. 1990 (September). *Finding the Rx for Managing Medical Wastes.* Washington, D.C.: GPO.

U.S. Environmental Protection Agency: Clean Air Scientific Advisory Committee. 1981 (September). *Setting Ambient Air Quality Standards: Improving the Process.* A Report of the Clean Air Scientific Advisory Committee, Science Advisory Board. Washington, D.C.: U.S. E.P.A.

U.S. General Accounting Office. 1983 (Aug. 26). *Delays in EPA's Regulations of Hazardous Air Pollutants.* Washington, D.C.: GPO.

———. 1984 (Sept. 27). *Status of EPA's Air Quality Standards for Carbon Monoxide.* Washington, D.C.: GPO.

———. 1989 (May). *Infectious Waste: Federal Health Care Facilities' Handling and Disposal Practices.* Washington, D.C.: GPO.

———. 1990 (March). *Medical Waste Regulation: Health and Environmental Risks Need to be Fully Assessed.* Washington, D.C.: GPO.

U.S. Congress. House. 1983 (November 7). Hearings before the Subcommittee on Oversight and Investigations of the Committee on Energy and Commerce on EPA's Air Pollution Control Program, 98th Congress, 1st Session. Washington, D.C.: GPO.

———. 1987 (October 27). Hearing Before the Subcommittee on Transportation, Tourism and Hazardous Materials of the Energy and Commerce Committee to Review Issues Relating to the Regulation of Solid Waste Management and Transportation, Focusing on Infectious Medical Wastes. Washington, D.C.: GPO.

———. 1988 (August 9). Hearing Before the Subcommittee on Regulation and Business Opportunities and Energy of the Committee on Small Business on Health Hazards Posed in the Generation, Handling and Disposal of Infectious Wastes. Washington, D.C.: GPO.

———. 1989 (July 25). Hearing Before the Subcommittee on Regulation and Business Opportunities and Energy of the Committee on Small Business on Infectious Wastes, A One-Year Update on Practices, Policy and Public Protection. Washington, D.C.: GPO.

Wagner, Kathryn D. 1987. "The Misdirected Mandate: Environmental Policies and the Politics of Rulemaking" Ph.D. diss., Cornell University, Ithaca, New York.

5

Technology, Analysis, and Policy Leadership: Congress and Radioactive Waste

Michael E. Kraft

After decades of ignoring the growing problem of radioactive waste disposal, the U.S. Congress in late 1982 enacted the Nuclear Waste Policy Act. That act launched a complex, comprehensive, and long-term program for the permanent disposal of high-level wastes from commercial power plants in geologic repositories. For five years the Department of Energy (DOE) struggled unsuccessfully to implement the policy amid increasing public and state opposition. Congress tried once again to resolve the conflicts and set a new policy direction with the Nuclear Waste Policy Amendments Act in 1987, which focused all effort on a single site at Yucca Mountain, Nevada. However, its implementation has been no more successful, and the outlook remains decidedly bleak. Very likely, Congress will be forced to confront the issue for the third time in a decade, much to the dismay of the nuclear power industry and its supporters. They hope to spur a revival of nuclear energy in the 1990s that is dependent upon solving the nuclear waste problem.

This record of policy failure raises important questions about the nation's ability to govern technology, particularly about how a representative legislature like Congress makes difficult policy choices. Does Congress have the requisite institutional capacity, for example in the Office of Technology Assessment (OTA) and in the committee staffs, to understand and resolve controversies over highly technical issues such as the best method for isolation of dangerous radioactive wastes from the biosphere? Or over which geological sites are most suitable for a repository? Or over the procedures that Executive Branch departments such as DOE should follow to insure that decisions are based on competent scientific and engineering knowledge? Presumably Congress is better equipped to handle the value conflicts and political questions, such as how many repositories to require and how to distribute them geographically, how to reconcile environmental disputes, how much of a role to grant to the states, whether some form of compensation should be provided to host states, and what level and type of public participation is warranted. But how well does it do so? Most importantly, does Congress get the policy analysis it needs to make intelligent decisions about a long-term commitment to facilities few people want nearby and whose environmental effects are highly uncertain but potentially severe?

In this chapter I examine the way Congress dealt with both the technical and political issues of nuclear waste over the past decade. I focus on how congressional policy makers have tried to meet two objectives often thought to be in conflict: 1) to design a technically sound repository siting program, and 2) to respond to public concerns over the health and environmental risks posed by such repositories. I argue that the two objectives are not necessarily at odds, and, indeed, that achievement of the second is a prerequisite for the first. It is only through a process of full policy legitimation—or democratic consensus building—that important political and technical issues, including potential obstacles to implementation, are likely to be understood and addressed effectively.

Radioactive Waste as a Policy Problem

The need for high-level radioactive waste disposal derives chiefly from the federal government's decision over thirty-five years ago to promote and heavily subsidize the commercial use of atomic energy. Since the first nuclear reactor was put on line at Shippingsport, Pennsylvania in late 1957, high-level wastes (inevitable byproducts of fission reaction) have been accumulating in water-filled basins at reactor sites around the nation (U.S. OTA 1985, 21–36). That method is unsuitable for long-term storage, and some utilities have exhausted, or soon will, their on-site storage capacity. Most of the wastes to be stored in the planned repository (as measured by radioactivity) come from these civilian power plants in the form of spent fuel rods. Hence the issue of nuclear waste disposal is intimately linked to the use of nuclear power.

Over the past decade, most analysts have assumed that commercial nuclear power had a dismal future, largely because of its high costs, stringent regulatory requirements, and public opposition (Morone and Woodhouse 1989; Campbell 1988). After 1978, no additional nuclear plants were ordered by the nation's utilities, and all of those ordered after 1974 were eventually cancelled. More recently, with growing concern about global warming caused by use of fossil fuels, the prospects for a renewed reactor construction program have improved (Nealey 1990). However, regardless of whether nuclear energy enjoys a revival in the 1990s, the United States must deal with the accumulated radioactive waste as well as that projected to be generated over the life of the 112 reactors now in operation. Other nations using nuclear power must do the same. These highly dangerous wastes must be isolated from the biosphere for up to 10,000 years. Assuming no new orders for reactors, DOE estimated in 1988 that the total amount of such high-level wastes by the year 2000 in the United States would be some 41,000 metric tons, and for the year 2020, nearly 87,000 metric tons (U.S. DOE 1988, 2). Should the U.S. embark on a new reactor program, as the Bush administration advocated in 1991 as part of its national energy policy, the total amount of waste would greatly exceed these projections.

Where should these wastes be placed? Over the years many alternatives have been discussed, including disposal in mined geologic repositories in several different media, subseabed or ice sheet disposal, deep hole disposal, island disposal, ejection into space, and long-term storage in various kinds of surface or sub-surface containers (U.S. OTA 1985). The first is clearly the most extensively studied alternative and the one favored by the scientific community. It involves a system of natural and engineered barriers that are designed to either prevent or greatly limit the escape of radiation, especially into groundwater. Technical questions abound, although few are considered to be insurmountable given sufficient time. Nevertheless, considerable uncertainly necessarily characterizes any projections of geological stability over a period of thousands of years. Nearly all policy analyses since the mid-1970s have focused on geologic disposal, although so-called Monitored Retrievable Storage (MRS) systems have been assessed as well.

The characteristics of radioactive waste disposal affect its politics in some predictable ways. Nuclear waste is at the top of the list of modern technological risks people most fear, making public perception of the risks important even when they differ dramatically from expert judgment (Slovic 1987; Kraft 1991a). No nation has yet built a repository; thus this is an untried and unproven technology. Providing assurances to the public of repository safety is difficult, particularly in a time of heightened environmental concern and distrust of government agencies.

In addition, siting a repository creates distributional issues of several kinds. Risks will be concentrated on the host community and state while diffuse and largely invisible benefits are provided to the population at large. Many analysts question the fairness of imposing such risks on an unwilling population. Politically, the need to locate a facility somewhere creates incentives for those near a possible site—and their state governments—to oppose the repository, and for their congressional representatives to try to insure it goes into someone else's district or state. There is little reason for residents of other states or their representatives to accord the issue nearly as much weight. Risks are also passed on to future generations, raising questions about intergenerational equity that have gone largely unanswered to date.

As we turn to the process by which nuclear waste moved to the political agenda and the 1982 and 1987 policies were formulated, special attention will be given to how well these aspects of the problem have been addressed by policy analysis. The analysis in question comes from several sources, including Executive Branch agencies and task forces, the OTA, and independent scientific organizations, over more than ten years. It is also important to ask whether the analysis reaching members of Congress provided clear, understandable, and timely answers to the questions members asked, or should have asked, and thus aided their decision making.

Evolution of the Nuclear Waste Policy Agenda

Enactment of a nuclear waste policy in the United States is surprisingly recent given civilian and military use of nuclear power for over forty years. From the dawn of the nuclear age in the 1940s until well into the 1970s, the federal government and the nuclear industry considered waste disposal to be a manageable and noncontroversial technical problem. Those beliefs, in combination with a highly favorable political climate, kept waste issues off the political agenda even though civilian use of nuclear energy, and thus waste generation, grew appreciably during this period. Contributing to the low saliency of the issues was dominance of the atomic energy agenda by a closely-knit nuclear subgovernment consisting of the nuclear industry, the Atomic Energy Commission (AEC), and the congressional Joint Committee on Atomic Energy (Temples 1980). They shared an optimistic view of nuclear power and were unconcerned about the waste issue (Rosa and Freudenburg 1992; Jacob 1990).

In 1955, the AEC finally asked the National Academy of Sciences to study nuclear waste. The Academy reported two years latter that the existing disposal practices were unsafe over the long term, and it recommended geological disposal in salt mines or repositories. However, the low priority of waste disposal continued. In retrospect, the AEC lost an important opportunity to devise a long-term solution under the highly favorable political climate for nuclear energy in the 1950s and 1960s.

The consensual and relatively closed politics of nuclear energy began to change in the late 1960s, and eventually made the politics of nuclear waste far more controversial. A new political climate emphasized environmental, health, and safety concerns, symbolized by passage in late 1969 of the National Environmental Policy Act (NEPA). Public fear of the risks associated with nuclear power and waste grew, due in part to problems of leakage at some military storage facilities (e.g., in Hanford, Washington) and difficulties with early efforts to locate a waste repository in an old salt mine near Lyons, Kansas. An explicitly antinuclear movement opposed to nuclear power arose as well. Moreover, public distrust of government increased sharply, posing special problems for secretive and technocratic agencies such as the AEC and its successors which adapted poorly to new expectations of open decision making.

During the early 1970s, the AEC under James Schlesinger's leadership announced plans to build a retrievable surface storage facility "to allow time to resolve the uncertainties" of nuclear waste disposal (Carter 1989, 7). Environmental groups objected to separating the issue of nuclear power from waste disposal in this way, and the Energy Research and Development Administration (ERDA), the successor to the AEC, dropped the idea. Instead, by the mid-1970s, ERDA announced several plans, one quite ambitious, for a nationwide search for sites for permanent geologic repositories, arousing strong protests from the states. About the same time, the California legislature tied the construction of new reactors to the availability of a proven, permanent waste disposal method. Other states followed its lead, sending a message to the nuclear industry that it must solve the waste disposal problem or face the decline of nuclear power. In addition, the Carter administration's decision to indefinitely postpone spent fuel reprocessing meant that the total amount of waste would rise sharply.

Formulating a Flawed Policy:
The Nuclear Waste Policy Act of 1982

The combined effect of these changes shifted nuclear waste issues from the back to the front burner. Many more participants sought to shape waste policy, and these new policy actors represented a far more diversified set of interests than in the old days of the highly consensual nuclear subgovernment. Environmental risks and health concerns in particular were propelled to new heights as environmental and other public interest groups achieved significant political influence. These new political forces were evident when DOE (which succeeded ERDA in 1977) embraced the deep geologic repository option. President Jimmy Carter called instead for a comprehensive study of the nuclear waste question, beginning a more formal process of policy formulation.

Executive Analysis and Decision Making

As often the case in the Carter presidency, policy formulation was delegated largely to an ad hoc task force, in this case a high-level interagency Nuclear Waste Management Task Force, which came to be called the Interagency Review Group (IRG). Consisting of 13 federal departments and agencies, the chief players were DOE, the Interior Department, the White House Office of Science and Technology Policy (OSTP), and the Council on Environmental Quality (CEQ). There is no doubt that Carter intended to give greater voice to environmental interests, and indeed the CEQ, the U.S. Geological Survey (USGS), and OSTP all battled with DOE on a range of issues (Greenwood 1982).

The IRG's study was both thorough on the technical aspects and attentive to many of the political issues noted above. It also began the equally important process of policy legitimation. The latter was evident in the circulation of an October 1978 draft report, 15,000 copies of which were distributed. The IRG held public hearings in Washington and across the country, arranged meetings with interest groups, and solicited public comments on the report. A final report was issued in March 1979, and could best be described as a discussion of objectives and procedures for developing such a waste plan, including analysis of political and administrative issues such as the role of the states and public participation (IRG 1979).

The IRG concluded that knowledge of the technical and scientific aspects of repository siting and construction was deficient, but could be improved as the process developed. It recommended that the nation study and screen multiple sites in different geologic media, choose the best that could survive NRC's licensing criteria for nuclear installations, and construct the repository itself using both natural and engineered barriers for greater assurance of containment. The report reflected the conviction of experts in DOE that geologic disposal was preferable to the alternatives. This technology of permanent, irretrievable disposal was also preferred as a way to limit the proliferation of nuclear weapons—the plutonium in spent fuel would be inaccessible. In addition, as Luther Carter (1989) has noted, the IRG emphasis on comparison of possible sites was consistent with NEPA's philosophy of conducting environmental assessments and considering alternative policy actions. It also reflected the new political climate of the late 1970s. The IRG members believed there was no imminent danger from continued storage of nuclear waste, and therefore that a wide survey of sites was both feasible and helpful to decision making—and more important than construction of a repository at an early date. For the same reasons, the IRG endorsed the concept of "consultation and concurrence" with states and Indian tribes to emphasize the importance of public information and comment. Much of this procedural philosophy was endorsed by Congress in 1982—and reversed in part in 1987.

The IRG's technical analysis was well grounded in geologic science. Its call for an elaborate site evaluation process and public participation did suggest, however, that its own estimate of having a licensed repository by the early 1990s was unrealistic. Geological Survey scientists expressed strong doubts during the IRG process and in later congressional hearings that sufficient earth science information would be available by the proposed deadlines to allow DOE to make objective decisions about site suitability (U.S. Congress 1981). The IRG also underestimated the likelihood of intense public and state opposition to the siting effort, although it should have been alert to this possibility given the states' hostile reaction to ERDA's disposal plans in the mid-1970s.

Finally, in a move that many later came to regret, the IRG considered and rejected arguments that, in order to enhance public credibility in the effort, an agency other than DOE should be placed in control of the repository program (Greenwood 1982). Both DOE and Congress ignored similar advice in a 1982 OTA study (discussed below). They did so again in 1984, when a DOE advisory panel recommended that the Office of Civilian Radioactive Waste Management (OCRWM) created by the 1982 act be replaced by a public Federal Corporation for Waste Management chartered by Congress to improve credibility, internal flexibility, and cost effectiveness (Colglazier and Langum 1988, 338).

President Carter's recommendations to Congress were delayed by disputes between DOE and the CEQ, and other factors, and they were announced only in February 1980. In the interim, the Three Mile Island accident in 1979 sharply raised public doubts about nuclear power (Rosa and Freudenburg 1992) that would stimulate dissent on Capitol Hill. In the main, Carter endorsed the IRG's plan for a broad, nationwide search for a repository. He also stressed that the program should include the "fullest possible disclosure to and participation by the public and technical community" (Carter 1980, 220–24).

Congress Enters the Fray

Although the IRG process and White House decision making represented a comprehensive executive formulation of nuclear waste policy, Congress acted independently on the issue in the 96th Congress (1979–80) while the Carter study group was still at work. The divisions that would characterize congressional debate on the subject for the next decade were already evident. The Senate had J. Bennett Johnston (D., Louisiana), chair (and from 1981 to 1986, ranking minority member) of the Energy and Natural Resources Committee, playing the lead role. He worked closely with ranking minority member James A. McClure of Idaho, who served as committee chair from 1981 to 1986. Johnston was skeptical of the IRG's preference for the geologic disposal option, and he wanted Congress to commit itself to two key actions. One was a guarantee that the federal government would accept the industry's spent fuel to relieve them of that burden and "solve" their waste problem. He also wanted rapid construction of temporary storage facilities, an MRS system, to speed the process along. He believed surface or near-surface storage offered "the fewest uncertainties and the most flexibility" compared to other alternatives, particularly geologic disposal (Cooper 1989, 6). To insure that the MRS system would not be sacrificed to a geologic repository, Johnston called for limited application of NEPA's provisions that alternatives be considered.

Johnston's opposite number in the House was Representative Morris K. Udall (D., Arizona), chair of the Interior and Insular Affairs Committee. Udall shared the environmentalists' distrust of a surface facility, and he preferred a geologic repository, full compliance with NEPA, licensing by the NRC, and strong state and public roles in decision making (Congressional Quarterly 1980). Thus he broadly endorsed the IRG's recommendations. In particular, he was concerned that too rapid a repository siting schedule would arouse the public and states, and undermine political support for the geologic disposal policy. He believed that a thorough investigation of multiple sites as recommended by Carter's IRG was the best way to meet the two key objectives: to identify technically suitable sites and to gain public acceptance of the repository.

Congress nearly passed a comprehensive policy in 1980, before Carter's recommendation arrived, but jurisdictional disputes among House committees and sharp House-Senate differences over the MRS short-term solution prevented final action (Congressional Quarterly 1980, 494). DOE lobbyists did not press hard on the legislation, presumably because they were awaiting the final IRG report and presidential recommendations. Congress was able to agree on a *low-level* radioactive waste bill, which it enacted in December 1980.

Most of the issues debated in the 96th Congress reappeared in the 97th Congress in 1981 and 1982. The Senate Energy and Natural Resources Committee cooperated with the Environment and Public Works Committee, which shared jurisdiction on some aspects of the policy. As before, the Senate acted quickly and approved its bill on a 69 to 9 vote following heavy lobbying by the Reagan administration and the nuclear industry (Congressional Quarterly 1982, 306). The Senate's version was a "fast-track bill" with an ambitious schedule. It called for selection of three potential sites by 1984, selection of one of them by 1986, and operation of the facility by the mid-1990s. It provided for NRC licensing for both construction and operation of the facility, but again restricted the application of NEPA to the decisions. Environmentalists were disturbed by the timetable as well as by the limited role given to the states, and they were aided by a last-minute filibuster threat by Senator William Proxmire (D.,

Wisconsin) that resulted in a provision for a state veto on siting (subject to congressional override). To avoid placing all the waste in a single state, and to distribute the political risks geographically, Senators Slade Gorton and Henry Jackson of Washington proposed setting a tonnage limit on the repository. The Senate agreed, an action interpreted later as a congressional mandate for two repositories, one in the West and one in the East.

The House had a more difficult time passing a bill given the jurisdictional battles among the various committees with some authority over this legislation. By one count seven committees were involved, with Energy and Commerce, Interior, and Science and Technology playing important roles. Key issues included whether to provide for interim federal storage of spent fuel (favored by the nuclear industry and opposed by environmentalists), what to do about military waste, and what authority to grant to the states. The House insisted on elaborate procedural guarantees out of belief that only careful, multiple site evaluation offered assurance of a safe repository. The electric utilities and nuclear industry were so eager to get a bill passed that they agreed to let the details be worked out during implementation. DOE projected that the cost of multiple assessments would be minimal, and thus overall program costs were not a major issue at this time (Cooper 1989, 10). A compromise bill, largely reflecting the Senate's Johnston-McClure proposal, was assembled in early December 1982 and enacted on the last day of the session. It was signed into law by President Ronald Reagan on January 7, 1983.

The issues raised in congressional debate confirm that Congress in the early 1980s was more representative of diverse interests on nuclear power and related questions than it was in the 1950s and 1960s. Advocates for nuclear power continued to have an influential voice, but they were joined by many new players, particularly environmentalists and the National Governors' Association, representing the states. It is clear as well that Congress succeeded to some extent in balancing these competing interests. This was evident in the NWPA's specifications on the state veto, public participation, and the call for a detailed DOE Mission Plan that would set out the procedures to be followed. The final act, however, was a brittle compromise that signaled deep congressional ambivalence about the repository issue. Members recognized the need to construct a repository, but they were hardly eager to have one located in their states. They authorized a procedure intended to produce a repository by 1998 (a highly ambitious schedule), yet they also allowed reasonably full state and public involvement (now called "consultation and cooperation") that was potentially inconsistent with that goal. Moreover, they provided no compensation to the host states nor any incentive for state cooperation with DOE, a significant flaw in light of later events.

These policy design weaknesses meant there was no firm consensus on nuclear waste policy to guide the act's implementation. As discussed below, the act's failure and the outcome of Congress's 1987 redirection of NWPA pointed to continuing divisions along the same lines. An appropriate question, therefore, is to what extent did the various policy analyses available to Congress help set policy goals and means and resolve these differences?

The Office of Technology Assessment Study

By the time Congress was fully engaged with the core issues in 1982, it had the benefit not only of the formal IRG report and Carter's legislative recommendations, but also a study prepared by the Office of Technology Assessment and released in April 1982 (U.S. OTA 1982). OTA entered the picture in an unusual way. Its assessment of federal policy for management of commercial high-level waste was prepared at the request of a House committee marginal to the major issues, the Committee on Merchant Marine and Fisheries, which was interested in the possibility of ocean disposal of waste. OTA later expanded its coverage of the issues when interest was expressed by other committees, including Johnston's Energy and Natural Resources Committee and Udall's Interior Committee. It presented its findings and recommendations both in committee testimony and in a summary report. According to OTA, it surveyed all major views on the nuclear waste question and then identified the policy elements that addressed the concerns expressed by these parties. It focused on those policy options that "appeared capable of securing the credibility, stability, and broad support essential to a successful waste management effort" (U.S. OTA 1985, 99).

Its resultant "integrated policy" stressed the need for a comprehensive geologic disposal policy that commanded broad support, embodied a "firm and conservative schedule" for the permanent repositories as well as a commitment to utilities to accept waste on a conservative date, and provided the requisite administrative and financial resources to meet those goals. Vital to its success, OTA said, was adequate institutional capacity to implement such an ambitious effort and an ability to build public acceptability of the facility and credibility for the implementing agency. In its own analysis in 1985 (OTA 1985, 101–111), OTA concluded that the NWPA contained most of the elements it identified in 1982 as essential. The key omissions were a conservative siting schedule that would provide assurances of safety and creation of an independent waste management agency. These were two of the most critical policy elements, and their omission contributed significantly to the act's failure. The OTA report observed also that DOE's Draft Mission Plan of 1984 fell short of its potential for providing a basis for "informed decisions" and for building credibility and acceptability of the waste disposal program.

There is no way to determine precisely what impact these various studies had on Congress collectively or on key members. The IRG and OTA recommendations were broadly consistent with what Congress ultimately approved in late 1982, one indicator that their analysis affected decision making in some important ways. A consensus emerged on the virtues of a geologic repository even if the MRS approach was never abandoned. Yet many commitments were more easily made on paper than achievable during implementation. A case in point is the repository siting schedule, which was later widely criticized. Another is the role accorded the states and the public in siting decisions, which led to such strident conflict with DOE that implementation was derailed by 1986.

Luther Carter (1987 and 1989) and other observers have described the congressional agreement on NWPA as a fragile compromise. Legislative compromises rarely have the clarity and consistency of objectives that policy analysts seek. That was the case with NWPA. Indeed, Carter characterized the final 1982 act as "too far-flung, too complex technically, too tortuous procedurally, and too controversial" to be implemented. DOE, he says, became "a Johnny Appleseed of dissension, walking through state after state planting political firebombs by identifying candidate repository sites" (Carter 1989, 12).

Implementing the 1982 Act

As DOE began implementation of the NWPA, it cited the congressionally imposed deadlines as justification for emphasizing a "fast track" timetable for site evaluation (Isaacs 1989, 4). The department issued draft siting guidelines and its Draft Mission Plan amid complaints that it was moving too quickly for sound scientific judgments to be made, as Geological Survey scientists had warned years earlier. The Reagan administration compounded these difficulties by delaying until May 1984 the nomination of a permanent director for OCRWM. By December 1984, DOE faced mounting public opposition as the department began focusing on particular repository sites.

On December 19, 1984, DOE ranked nine candidate sites in six states for the first of the two repositories authorized by NWPA, and it named the top three for site characterization: Deaf Smith County, Texas; Yucca Mountain, Nevada; and Hanford, Washington. All three states protested and eventually filed suit against DOE, challenging the legality of the designation process. Much of the dispute centered on the environmental assessments. Criticism, including that from the National Academy of Sciences (NAS) Board on Radioactive Waste Management and the Nuclear

Regulatory Commission, was directed especially at omissions and other deficiencies in the data used, the methodology for site evaluation and comparison, and DOE bias in the evaluation process (Clary and Kraft 1989; Colglazier and Langum 1988).

Similar controversies surrounded the Second Round siting process in the East. On January 16, 1986, DOE announced that 20 locations in seven states in the Upper Midwest or East had been selected for further study from its original list of 235 potential sites in 17 states. Once again DOE defended its choices in terms of the official siting guidelines. Public hearings in the potential host states in the spring of 1986, attended by thousands of individuals, representatives of environmental groups and Indian tribes, and state and local government officials, revealed massive public opposition to DOE's siting proposals. Both the hearings and public opinion surveys pointed to a widely shared perception that DOE lacked technical competence and credibility with the public (Kraft and Clary 1992; Dunlap, Kraft, and Rosa 1992). The Chernobyl nuclear accident in April 1986 did little to improve DOE relations with a public that already feared possible contamination by radioactive waste and that was becoming increasingly skeptical about nuclear power itself (Rosa and Freudenburg 1992).

Intense opposition of this kind in the East and fears of political repercussions for Republican Senate candidates led the Reagan White House in late May 1986 to put the siting process for the Eastern states on hold. Chances for successful implementation diminished further when the western states objected strenuously to what they viewed as inequitable treatment that left them vulnerable to imposition of a repository. As a result of these and other criticisms, Congress imposed a moratorium on the use of funds for exploratory work in the West, and in 1987 it considered how to revise the act.

Legislative Repair and Redirection: The Amendments Act of 1987

During 1987 some 20 different bills were introduced in an effort to amend the NWPA. Two quite different proposals emerged from the pack, one in the Senate and one in the House. Led by Morris Udall, the House preferred a continued moratorium on siting efforts and further study. Udall argued that "DOE blew it," declaring that "the program is in ruins" and that "the public and many of us in Congress have lost all faith in the integrity of the process." Thus he wanted to start over (Congressional Quarterly 1987, 310). He called for putting the selection process on hold for 18 months to allow a bipartisan commission within the legislative branch to thoroughly review the 1982 policy and DOE's implementation of it. The commission would examine DOE siting guidelines, environmental assessments, ranking methods, and consultation with the states, and even whether DOE should continue to run the program. Udall wanted OTA to conduct the study, but it declined to do so, citing insufficient staff resources. Another of Udall's proposals created a high-level negotiator (operating out of the executive office of the president) to work with the states to find a willing recipient for the repository. The Interior Committee approved these measures in late October, 1987.

In light of events of the past three years, the Udall approach may well have been a sensible way to go in 1987. However, Bennett Johnston was convinced such an approach would end all hope of continuing with the 1982 act. He also feared that the waste issue would be vulnerable to presidential election politics if Congress didn't settle matters in 1987. Thus he pressed hard for a sequential approach to site characterization that would focus on a single site at a time and thus facilitate a rapid repository siting schedule. He argued this would save considerable money (the costs of such study having risen sharply to an estimated $2 billion per site) as well as limit the scope of political conflict. Some of the savings (up to $100 million per year) would be offered to the potential host state as an incentive to cooperate. Johnston continued to urge action on an MRS facility as well (Cooper 1989).

This legislative package was formulated by professional staff on the Senate Energy and Natural Resources Committee independently of DOE, although the department (and utility companies) pleaded with them to help save the program (Cooper 1988). Like Johnston, DOE officials feared that a moratorium would lead to collapse of the whole repository effort (Nelson 1988). In a strategic use of scientific studies and testimony (Whiteman 1985), the committee held 11 hearings in 1987 to solicit views on program implementation and technical problems, and especially to create a "public record" that it believed documented the need for its approach (Cooper 1989, 20–21). That objective doubtless limited the extent to which the information provided would alter members' policy preferences. Nevertheless, they learned of genuine problems with the program. For example, the committee knew well that DOE's secretive culture and poor working relations with the states were key obstacles. "That's one of the things that came up heavily in our hearings," said one committee staff member (Meigs 1988). The hearings also included testimony by representatives from the Nuclear Regulatory Commission and the National Academy of Sciences indicating there was no technical reason to disqualify any of the three final sites in the West prior to characterization. Hence Johnston could argue that there was no justification on those grounds for a moratorium.

The Senate endorsed the Johnston proposal by a two to one margin, a testimony to his extraordinary skills in assembling a supportive coalition and his aggressive resistance to any moratorium on the siting process. By eliminating the eastern sites, he gained support from delegations of the 17 potential host states in that region. He also worked closely with senators from Tennessee (the candidate for the MRS facility) and those from two of the three final candidate states, Texas and Washington, to meet their objections. One committee staff member described Johnston's actions as "heavy hitting, hardball politics," and noted that he "virtually black-mailed" senators into supporting his position by threatening to reopen consideration of their states or to cut energy and water projects slated for their states (Berick 1988). A DOE official was more positive. "Johnston is an astute politician, and he knows how to marshall his forces and he knows how to split off the opposition" (Nelson 1988).

Johnston also adopted a three-way legislative strategy, which gave him a great advantage over Udall. Said one Senate staff member, "we had it wired, we had it many ways." Johnston developed a "stand alone" bill in his Energy Committee as well as two other bills, one tied to a budget reconciliation measure in the Budget Committee and one to an energy and water appropriations bill in the Appropriations Committee. He was a member of both committees, chaired the Energy Committee in 1987, and was prepared to pursue whichever avenue seemed most likely to give him a swift victory. He did encounter some opposition from the Senate Environment and Public Works Committee, which, like Udall, urged a delay in the program until the early 1990s. Committee members were irritated that Johnson had moved ahead without consulting with them (Cooper 1988), and they worked closely with environmentalists to draft their own measure. However, they began far too late in the year and had too little influence to alter the Senate's position. Johnston easily undercut the modest coalition being built for a moratorium.

Backers of the moratorium approach, including environmentalists, were far better represented in the House, with Udall joined by two influential senior members, John Dingell and Philip Sharp, in advocating the plan approved by the Interior Committee. On procedural grounds they opposed Johnston's use of the appropriation process for revising NWPA. They also opposed granting any compensation to the host state and continued to reject the MRS facility, which was a sharply contentious issue at this time. They persuaded Johnston to talk with them informally in a series of meetings in December 1987. Eventually, with intervention by the House speaker and majority leader (from Texas and Washington, respectively), Johnston agreed to compromise on the MRS issue and the House agreed to drop its moratorium proposal. In an unexpected move that caught Johnston by surprise, the House delegation decided to name directly the one western site to be considered: Yucca Mountain, Nevada. Hence the House concluded that the repository program should go forward despite the many problems encountered.

As for the explicit designation of the Yucca Mountain site (technically absent from the Senate bill, but implied in its siting criteria), members from Texas and Washington argued that the environmental risks of the Texas and Washington sites (contamination of groundwater and rivers) were greater, making the Nevada site preferable. Yet they also acknowledged that forces beyond their control were shifting the site selection process from a scientific to a political one. Rep. Al Swift (D., Washington) later called the plan a "goddamned outrage," and said "we've done it in a purely political process." Yet he justified his position in favor of it by saying he had to protect his constituents (Davis 1987, 3136; Congressional Quarterly 1987, 310). The Nevada delegation called the measure the "Screw Nevada Bill." But as representatives of a small state lacking senior members in Congress, they had little political influence to stop it.

Congress went on to approve a compromise package of amendments to the NWPA that directed DOE to evaluate Yucca Mountain and to cease consideration of the other two western sites. It also cancelled authorization for a second repository in the Midwest or East, nullified DOE's selection of Tennessee for an MRS facility, and authorized a special negotiator appointed by the president to seek a state or Indian tribe prepared to accept either the permanent repository or the temporary facility. Johnston's original proposal of $100 million a year for the host state was reduced to a far more modest $20 million—and with a stipulation that to get it Nevada would have to relinquish its right to veto its selection. Congress did authorize a temporary, surface facility, but not until construction of the permanent repository was licensed (PL 100–203; Cooper 1989). Critics from many quarters, including Nevada, complained that the act was destined to fail again, and that the procedures by which it was approved violated standards of both policy legitimation and ethics (Lemons, Malone, and Piasecki 1989; Lemons, Brown, and Varner 1990).

DOE Tries Again

The amendments act of 1987 appeared promising at first. DOE would deal with only a single site in one state, thus greatly narrowing political opposition. The other 49 states had little reason to criticize DOE or the program and thus risk reopening the issue of repository location (Cooper 1989). Much of this optimistic prognosis depended, however, on the willingness of Nevada to cooperate with DOE, and on DOE's competence to carry out the complex site characterization process.

As it turned out, three years of bitter relations between the state and DOE followed, which many attributed to Nevada's "wounded psyche" at being singled out for the repository. DOE, now under a new secretary, James D. Watkins, rejected its previous studies of Yucca Mountain, declaring in late 1989 that it would begin a new assessment. Nevada used such admissions and the federal funds made available for its own site review and socio-economic impact studies to challenge DOE at every opportunity. It soon became clear that few state officials believed cooperation with DOE was in their political self-interest, suggesting the flawed logic behind Congress's grant of financial compensation as a supposed inducement for cooperation by the host state. The result of all this was continued state opposition and policy gridlock (Isaacs 1989; U.S. DOE 1989). Relations between the state and DOE deteriorated steadily until the Nevada legislature in mid-1989 in effect vetoed placement of the repository at Yucca Mountain. As a consequence, state agencies refused to issue environmental permits essential for DOE excavation and testing at the site, and all site activity was brought to a halt. The state maintained this position even after a federal appeals court ruled in September 1990 that its veto was illegal. In October, DOE asked Congress to intervene to settle the matter.

Only in early 1990 did the Bush administration finally name a permanent director for OCRWM, a position critical to the program's success. Six months later, the nuclear waste negotiator called for in the 1987 act was appointed. Yet by early 1991, he was making little headway with the states and Indian tribes (Wald 1991). Despite congressional redirection of the program in 1987, prospects for the repository program seemed about as dim as they were when it collapsed in 1986. A National Research Council panel concluded in mid-1990 that the program was highly unlikely to succeed because of public fears and opposition and a high degree of inflexibility in the work schedule, technical specifications, and regulatory standards (National Research Council 1990). Similar critiques and recommendations were offered by other reputable scientists and organizations, suggesting the need for further congressional action to amend the repository plan or replace it.

Conclusions: Policy Leadership and Nuclear Waste

The development of nuclear waste policy in Congress says much about the way in which policy analysis influences decisions on major issues of science and technology policy. The case illustrates how even the best studies from executive branch agencies, OTA, or independent scientific organizations, in addition to analyses of failed policy implementation, may not settle all controversial issues. Policy analysis is intended to clarify alternative courses of action and their consequences and thereby inform political decision making. But it is not a substitute for the judgment of public officials. Often, of course, there are strong political pressures on policy makers that may result in their giving scant attention to the kind of questions analysts believe are critical to successful policy design and implementation. Or they may selectively perceive and use studies to reinforce their previously defined positions, as David Whiteman (1985) has found to be the case with many OTA studies. The final policy outcome is shaped by the quality and perceived utility of the available analyses as well as by the characteristics of the decision-making process itself. Policy leadership is an important ingredient of that process.

When policy makers disagree sharply about the wisdom of policy actions, we rely on a democratic process of policy legitimation to insure that appropriate evidence and arguments are considered. Our hope is that critically important information and perspectives are introduced into the legislative process and that the resulting policies are thereby improved. Policy leaders or entrepreneurs are crucial in this process. They keep informed about technical and political developments. They use their positions of respect and influence to organize forums that allow important new analyses to gain visibility. They also know how to take advantage of windows of opportunity such as a sudden shift in the political climate. In all these ways they help to connect knowledge to policy (Kingdon 1984).

In the present case, the ability of Congress to learn from DOE's aggressive and error-prone implementation of the 1982 act was short-circuited by Sen. Johnston's leadership style. He was a zealous advocate of his position, and he allowed few opportunities for those who disagreed with him to voice their concerns. To be sure, few members wanted to reopen the whole policy again and put themselves and their states at risk (Cooper 1989). Their fear was exploited by Johnston, who was convinced that only an MRS facility or characterization of a single site could work given public attitudes toward nuclear waste. In reaching that conclusion, he relied heavily on the views of writers such as Luther Carter (1987) and on the assurances of high-level DOE officials that Yucca Mountain was indeed the best site available.

One DOE official called Johnston's success a "masterful legislative coup," but others challenged that assessment. One key staff member deeply involved in Senate actions described him as "pushy" and "relent-less" in pursuit of his bill: "He was not afraid to try any procedure, any mechanism, to get this bill to the floor and voted on. And he didn't care whose toes he stepped on or what precedent he set." Aside from creating bitter feelings among colleagues in Congress, such determination was not conducive to imaginative redesign of the waste policy. Johnston's eagerness to complete congressional action in 1987 and get on with the repository program resulted in an amendments act that was no more successful than the original 1982 policy. It is at least conceivable that another style of policy leadership, perhaps closer to Rep. Udall's, would have yielded a more productive result.

Policy failure here is attributable in part then, to an unrepresentative, and sometimes hurried, legislative process that disregarded or downplayed significant questions about the state of scientific knowledge, the repository siting schedule, public attitudes toward the repository, citizen participation in siting decisions, ethics, and the capacity of DOE to administer the program. From the perspective of policy proponents, the price paid for these oversights was high. The department's credibility and public support plummeted and the program was thrown into disarray. The policy's critics had more reason to welcome these outcomes, for they gained the opportunity to challenge the scientific analyses as well as policy goals and means, and to shape the long-term direction of nuclear waste policy. One could detect those future directions in the recent report by the National Research Council (1990), among other sources (Whipple 1989; Cook, Emel, and Kasperson 1990; Kraft 1991b).

There are also encouraging aspects of this case that suggest a more positive view of the potential for policy analysis in Congress. There is clear evidence here that thorough and highly competent policy analyses were conducted and made widely available to the key actors both within Congress and outside of it. These range from the IRG report of 1979 to OTA's 1982 and 1985 studies, to a seemingly endless series of implementation assessments by the General Accounting Office (GAO) and oversight hearings by House and Senate committees. Congress was provided with abundant documentation of at least some of the scientific, administrative, and political issues of nuclear waste disposal, and many members did make considerable use of the information. The existence of those studies helps to explain why Congress has been as critical of DOE as it has, and why it mandated not only the original DOE Mission Plan for the program, but annual reports to Congress on progress and further needs. On the whole, then, the problem was not lack of analysis, but insufficient use of it at critical junctures.

One lesson we might draw from this case is that to be useful in settling such conflicts, the scope of policy analyses must be broadened to include thorough evaluation of political, social, and ethical issues. A concentration on assessment of technologies and environmental impacts, and decision and management science, will not be sufficient, and will tend to focus political debate on technical rather than political or institutional issues. In fact, DOE has long ignored social science studies in its repository program work (Bradbury 1989). Another conclusion is that a representative process of debate and deliberation facilitates criticism of deficient policy analysis and thereby promotes rational decision making. In the American political system we look to Congress to provide such representation of diverse societal interests. Generally it does this quite well, indeed, often to the point where the representation of conflicting policy outlooks creates political gridlock.

As Gary Bryner argues in Chapter 1, however, Congress faces a greater challenge in dealing with science and technology policy than it does for many other issues. In responding to problems such as nuclear waste and environmental protection, members need more help than they often get. Solutions surely lie in improving Congress's institutional capacity to anticipate and study such problems. Yet more and better policy analysis by itself is insufficient. The case of nuclear waste disposal strongly suggests the need to increase the motivation of policy makers to seek out and use such analysis. Only in that way can the nation succeed in managing technology through democratic politics.

Notes

This research was supported by a grant from the Everett McKinley Dirksen Congressional Leadership Research Center, whose assistance is gratefully acknowledged.

Bibliography

Berick, David. 1988. Interview with David Berick, staff of Senator Brock Adams. Washington, D.C., 7 June.

Bradbury, Judith. 1989. "The Use of Social Science Knowledge in Implementing the Nuclear Waste Policy Act." Ph.D. diss., University of Pittsburgh.

Campbell, John L. 1988. *Collapse of an Industry: Nuclear Power and the Contradictions of U.S. Policy.* Ithaca, N.Y.: Cornell University Press.

Carter, Luther J. 1987. *Nuclear Imperatives and Public Trust: Dealing with Radioactive Waste.* Washington, D.C.: Resources for the Future.

———. 1989. "Nuclear Waste Policy and Politics." *Forum for Applied Research and Public Policy* 4 (Fall): 5–18.

Carter, Jimmy. 1980. "Appendix A: Presidential Message and Fact Sheet of February 12, 1980." In *The Politics of Nuclear Waste,* edited by E. W. Colglazier.

Clary, Bruce B., and Michael E. Kraft. 1989. "Environmental Assessment, Science, and Policy Failure: The Politics of Nuclear Waste Disposal." In *Policy Through Impact Assessment,* edited by Robert V. Bartlett. Westport, Conn.: Greenwood Press.

Colglazier, E. W., ed. 1982. *The Politics of Nuclear Waste.* Elmsford, New York: Pergamon.

Colglazier, E. W., and R. B. Langum. 1988. "Policy Conflicts in the Process of Siting Nuclear Waste Repositories." *Annual Review of Energy* 13: 317–57.

Congressional Quarterly. 1980. *Congressional Quarterly Almanac 1980*. Washington, D.C.: Congressional Quarterly, Inc.

———. 1982. *Congressional Quarterly Almanac 1982*. Washington, D.C.: Congressional Quarterly, Inc.

———. 1987. *Congressional Quarterly Almanac 1987*. Washington, D.C.: Congressional Quarterly, Inc.

Cook, Brian J., Jacque L. Emel, and Roger E. Kasperson. 1990. "Organizing and Managing Radioactive Waste Disposal as an Experiment." *Journal of Policy Analysis and Management* 9 (Summer): 339–366.

Cooper, Benjamin S. 1988. Interview with Benjamin S. Cooper, professional staff member, Senate Committee on Energy and Natural Resource, Washington, D.C., 6 June.

———. 1989. "The Nuclear Waste Policy Amendments Act of 1987." Paper presented at the annual meeting of the American Association for the Advancement of Science, San Francisco, January.

Davis, Joseph A. 1987. "Nevada to Get Nuclear Waste; Everyone Else 'Off the Hook.'" *Congressional Quarterly Weekly Report*, 19 December, 3136–38.

Dunlap, Riley E., Michael E. Kraft, and Eugene A. Rosa eds. Forthcoming 1992. *Public Opinion and Nuclear Waste: Citizens View Repository Siting*. Durham, N.C.: Duke University Press.

Greenwood, Ted. 1982. "Nuclear Waste Management in the United States." In *The Politics of Nuclear Waste*, edited by Colglazier.

Interagency Review Group on Nuclear Waste Management. 1979. *Report to the President by the Interagency Review Group on Nuclear Waste Management*, Washington D.C.: Department of Energy, TI 29442, March.

Isaacs, Thomas H. 1989. "Implementing the Nuclear Waste Policy Act and Its Amendments." Paper presented at the annual meeting of the American Association for the Advancement of Science, San Francisco, January.

Jacob, Gerald. 1990. *Site Unseen: The Politics of Siting a Nuclear Waste Repository*. Pittsburgh: University of Pittsburgh Press.

Kingdon, John, W. 1984. *Agendas, Alternatives, and Public Policies*. Boston: Little, Brown.

Kraft, Michael E. 1991a. "Risk Perception and the Politics of Citizen Participation: The Case of Radioactive Waste Management." In *Analysis, Communication, and Perception of Risk*, edited by B. John Garrick and Willard C. Gekler. New York: Plenum.

———. 1991b. "Public and State Response to High-Level Nuclear Waste Disposal: Learning from Policy Failure." *Policy Studies Review* 10 (Fall), forthcoming.

Kraft, Michael E., and Bruce B. Clary 1992. "Public Testimony in Nuclear Waste Repository Hearings: A Content Analysis." In *Public Opinion and Nuclear Waste*, edited by Dunlap, Kraft, and Rosa.

Lemons, John, Charles Malone, and Bruce Piasecki 1989. "America's High-Level Nuclear Waste Repository: A Case Study of Environmental Science and Public Policy." *International Journal of Environmental Studies* 34: 25–42.

Lemons, John, Donald A. Brown, and Gary E. Varner. "Congress, Consistency, and Environmental Law: Nuclear Waste at Yucca Mountain, Nevada." *Environmental Ethics* 12 (Winter 1990): 311–27.

Meigs, Marilyn 1988. Interview with Marily Meigs, minority staff, Senate Committee on Energy and Natural Resources, Washington, D.C., 6 June.

Morone, Joseph G., and Edward J. Woodhouse 1989. *The Demise of Nuclear Energy? Lessons for Democratic Control of Technology.* New Haven: Yale University Press.

National Research Council. 1990. *Rethinking High-Level Radioactive Waste Disposal: A Position Statement of the Board on Radioactive Waste Management.* Washington, D.C.: National Academy Press, July.

Nealey, Stanley M. 1990. *Nuclear Power Development: Prospects in the 1990s.* Columbus, Ohio: Battelle Press.

Nelson, Richard. 1988. Interview with Richard Nelson, Director of Congressional Relations, Office of Civilian Radioactive Waste Management, Department of Energy, Washington, D.C., 3 June.

Rosa, Eugene A., and William R. Freudenburg 1992. "The Historical Development of Public Reactions to Nuclear Power: Implications for Nuclear Waste Policy." In *Public Opinion and Nuclear Waste*, edited by Dunlap, Kraft, and Rosa. Durham, N.C.: Duke University Press.

Slovic, Paul. 1987. "Perception of Risk." *Science* 236:280–85.

Temples, James R. 1980. "The Politics of Nuclear Power: A Subgovernment in Transition." *Political Science Quarterly* 95:239–260.

U.S. Congress. Senate. 1981. Committee on Energy and Natural Resources and Environment and Public Works Committee. *Joint Report on S1662,* 97th Cong., 1st sess., 30 November.

U.S. Department of Energy. 1988. *Commercial Nuclear Power 1988: Prospects for the United States and the World.* Washington, D.C.: Energy Information Administration, U.S. Department of Energy, 21 September.

———. 1989. *Report to Congress on Reassessment of the Civilian Radioactive Waste Management Program.* Washington, D.C.: Office of Civilian Radioactive Waste Management, 29 November.

U.S. Office of Technology Assessment. 1982. *Managing Commercial High-Level Radioactive Wastes: Summary.* Washington, D.C.: Government Printing Office.

———. 1985. *Managing the Nation's Commercial High-Level Radioactive Waste.* Washington, D.C.: Government Printing Office.

Wald, Matthew L. 1991. "Hired to Negotiate, but Shunned by All." *New York Times,* 13 February, 14.

Whipple, Chris. 1989. "Reinventing Nuclear Waste Management: Why 'Getting It Right the First Time' Won't Work." *Waste Management '89.* Tucson: University of Arizona.

Whiteman, David. June 1985. "The Fate of Policy Analysis in Congressional Decision Making: Three Types of Use in Committees." *Western Political Quarterly* 38: 294–311.

6

Congressional Decision Making and Long-Term Technological Development: The Case of Nuclear Fusion

W. D. Kay

Introduction

The U.S. program to develop controlled nuclear fusion as a source of useful energy represents an extreme case of a classic investment decision. It has been a highly expensive project, costing over $300 million each of the last 15 years and over $5.4 billion since 1951 (see Appendix B).[1] It is decidedly long term: even the most optimistic projections do no foresee working fusion reactors before the year 2030 (U.S. House of Representatives 1988, 30–31). Finally, it is extremely risky (in the investment sense), in that neither the technology nor even the basic physics of the fusion process are well understood.

On the other hand, the benefits of fusion power could be literally incalculable. In theory, it is safer than nuclear fission and less harmful to the environment than the burning of coal. Its fuel supply is thought to be, practically speaking, limitless: according to its supporters, the fusion process could produce the energy equivalent of 300 gallons of gasoline from just one gallon of seawater (U.S. OTA 1981, 29–30). Even more dramatic is the recent declaration by Edson C. Brolin, Acting Director of the Princeton Plasma Physics Laboratory, that "the top two inches of Lake Erie contains more energy than all the world's oil supplies" (Carey 1990, 62). In other words, the development of fusion technology promises nothing less than a complete and final solution to society's energy needs.

Although the costs and benefits of this particular investment decision may seem highly exaggerated, they are by no means atypical. The Strategic Defense Initiative, *Freedom* space station, and proposed Human Exploration Initiative of the Moon and Mars provide other examples of projects that are extremely expensive, have exceedingly long development times, and whose success depends upon the eventual resolution of present-day scientific and technological unknowns. There is every reason to expect that such programs will make up a significant portion of the U.S. R&D agenda for some time to come.

Because of its role in the budget process, it inevitably falls to Congress to make not only an initial "investment decision," but also any subsequent determination as to whether some ongoing project should be modified, delayed or even cancelled. Clearly, this is a challenging task. Successful development of these technologies requires a commitment of billions of dollars over a period of several years, or even decades, a fact which must be weighed against other budget priorities. In addition, because of their longer development times and inherent scientific uncertainties, projects cannot be realistically evaluated until they are already well under way. Stated simply, contemporary R&D policy calls for legislators to make increasingly important decisions with progressively less information.

This chapter will examine how Congress deals with such complex, expensive, and long-term technological projects, drawing primarily upon the experiences of the U.S. fusion energy program. Controlled nuclear fusion provides a useful case study for two reasons. First, as the brief description above suggests, it is characteristic of the types of technologies that are now beginning to crowd the Congressional R&D agenda. Secondly, it is among the oldest of these programs (dating back to the early 1950s), and thus provides an opportunity to observe how Congress manages such technologies over an extended period of time.

Nuclear Fusion Technology

Unlike the more familiar technique of nuclear fission, which involves "splitting" the nucleus of a single heavy atom (such as uranium or plutonium), fusion produces energy by "fusing" the nuclei of two light atoms, like hydrogen or helium (Glasstone 1980; Horgan 1989, 25–28; Niu 1990; U.S. OTA ch 2, 4). Fusion reactions are essentially what causes stars to "burn": their intense gravitational fields overcome the natural repulsion of subatomic particles, forcing the hydrogen atoms to fuse into helium while releasing excess energy in the form of light and heat (as well as various other types of radiation). The same process occurs in an "uncontrolled" form during the detonation of a hydrogen bomb.

There are, at present, two techniques potentially capable of achieving this effect on a controlled basis artificially. *Magnetic Fusion Energy* (MFE) fuses atoms by superheating hydrogen isotopes to temperatures in excess of 100 million degrees Celsius (the interior of the sun is estimated at approximately 16 million degrees). This superheated fuel (called *plasma*) must be kept at a very high density and temperature, confined within powerful magnetic fields. The result is somewhat like that of a low-gravity "sun."

Most of the current fusion energy research is of this type. The U.S. MFE program is managed by the Department of Energy's (DOE) office of Fusion Energy, and receives around $300 million annually. The governments of the USSR, Japan, and European Community (EC) also maintain active MFE research programs.

Inertial Confinement Fusion (ICF) works by bombarding a fuel pellet containing hydrogen isotopes with a high-intensity laser or particle beam, with the resulting implosion forcing a fusion reaction. Just as MFE simulates the nuclear processes of a sun, ICF resembles a miniature hydrogen bomb. U.S. inertial confinement research (which began in 1963) has therefore been limited almost exclusively to nuclear weapons simulations (and is classified). As of FY 1989, around 97% of ICF research (amounting to approximately $163 million) has been funded through DOE's Office of Defense Programs (National Research Council 1990, 4). As will be seen later, the technology's potential uses as an energy source have only recently received serious consideration. Consequently, most of the following discussion will deal with the magnetic confinement approach.

Several MFE devices have been developed, but one machine in particular, called a *tokamak*, has become the primary focus of current fusion efforts. Invented in the Soviet Union in the late 1960s (the name is a Russian acronym), this donut-shaped fusion device ranks as one of the most important innovations in the history of fusion technology. Most of the world's largest reactors—the Tokamak Fusion Test Reactor (TFTR) at Princeton, the European Community's Joint European Torus (JET), and Japan's JT-60—utilize this design.

Despite impressive advances over the last 40 years, in which the energy release from reactions has increased nearly 20 billion times (U.S. Congress, House 1988, 32), controlled nuclear fusion is still a very long way from demonstrating its feasibility as an energy source. No fusion reaction has yet "broken even," that is, produced as much energy as it has consumed. The magnetic fusion community generally expects, however, that breakeven will be achieved sometime in this decade.

Even though breakeven represents a historic scientific milestone, it is still order of magnitude below plasma *ignition*, in which the plasma becomes both hot and dense enough to create a self-sustaining "burn" without the addition of power from the outside. The most recent DOE plan calls for the construction of a reactor, the Compact Ignition Tokamak (CIT), that will achieve this effect allowing scientist to study the behavior of ignited plasma. The Department hopes to begin construction of the CIT during the 1990s, with operation beginning before the end of the century. Construction costs are currently estimated at $800–900 million (Crawford 1990, 119).

Even after (or if) this step is reached in a laboratory setting, developing a workable fusion technology requires the testing of engineering components and configurations. Under current planning, this step will be undertaken jointly by a consortium of nations. In 1987, the U.S. joined with the USSR, the EC, and Japan to begin conceptual designs of the International Thermonuclear Experimental Reactor (ITER), which will provide the first tests of MFE's engineering capabilities. The technology phase of this program is tentatively scheduled to begin around 2010. The total cost for ITER may reach as high as $4 billion, $1.6 billion of which would come from the U.S. (U.S. Department of Energy 1990, 11).

Determining whether or not the technology can serve as a useful source of energy, however, will require still one more step. Using the experience gained from ITER, the U.S. will then construct a fusion demonstration facility (DEMO). According to DOE's present timetable, DEMO will begin operation in 2020, producing "net electric power" and providing "the basis for fusion's commercialization" (U.S. Department of Energy 1990, 6). Only after significant operating experiences with such a device can it known with any degree of certainty whether this form of power generation is both technically and commercially viable (and the latter criterion, of course, will be dependent upon the status of competing energy technologies, consumer demand, etc.).

Thus, barring some major scientific or technological breakthrough, the development of practical, controlled fusion energy is still at least 30 years and three "reactor-generations" away. There is reason to believe, however, that even this schedule may be overly optimistic. To begin with, as will be seen below, MFE's timetable has already undergone numerous revisions over the last 40 years, and will almost certainly be revised again.

Secondly, serious scientific questions remain unanswered. The physics of fusion energy, particularly regarding the behavior of confined plasma, is still not well understood. Knowledge of plasma behavior is largely empirical, that is, derived almost exclusively from data obtained during reactor operation. There is not at present any sort of physical covering law which can predict or explain how confined plasma reacts to different conditions. Recent experiments on the TFTR, for example, have found that as an MFE reaction approaches 90 percent of the break-even point, the power output drops off. This phenomenon, referred to in the literature as *plasma transport*, cannot as yet be explained. It is this problem, more than any other, which has slowed recent progress of the MFE program (U.S. Department of Energy, *Report*, 1989).[2]

In addition, the technical problems associated with plasma confinement have proven to be far more complicated than originally believed. As suggested above, proper confinement is crucial for achieving a sustained fusion reaction. The superheated fuel must be kept away from the relatively cooler surfaces of the reactor interior while being maintained at a very high density. This has been a major technical obstacle for most of the program's history, and has still not been completely resolved.

Finally, some controversy has arisen concerning reactor fuel. The ideal fusion reactor would operate on pure hydrogen, the most plentiful element in existence. In practice, the most promising results come from reactions utilizing hydrogen isotopes. Some scientists have envisioned a process whereby the nuclei two deuterium atoms (a hydrogen isotope that occurs naturally in sea water) are fused (called a D-D reaction). By far the greatest energy yields, however, will come from fuel that combines deuterium with tritium, another isotope, to form a D-T reaction. It has been estimated that a single D-T reaction is a 4 million times more energetic than the burning of an equal amount of carbon (U.S. OTA 1986, 31).

There are two problems with using tritium as fuel, however, both of which seriously undermine some of the technology's potential advantages. First of all, tritium is rarely found in nature[3] and must therefore be manufactured (as the U.S. presently does for the construction of nuclear weapons). Secondly, it is radioactive, a fact which greatly complicates reactor design and operation, and raises significant environmental and safety issues (U.S. OTA 1986).[4]

In short, although magnetic confinement represents the most advanced approach to the production of fusion energy, and has already come a very long way, the technology remains, relatively speaking, in its infancy. Only after the next century is well underway will the actual benefits of nuclear fusion even be known, much less realized. In the meantime, Congress must weigh these scientific and economic uncertainties against the certain—not to mention heavy—budgetary requirements of the program.

Fusion Policy in the Congress

Amid great optimism, the U.S. MFE program was formally launched in 1951 (Bromberg 1981; National Research Council 1989, ch. 2; U.S. OTA 1986, ch. 3). Under the auspices of the Atomic Energy Commission (AEC), research centers were organized at Oak Ridge, Los Alamos, and Lawrence Livermore laboratories. Federal funding also helped establish fusion research centers at a number of universities, most notably the Princeton Plasma Physics Laboratory (PPPL). Compared to present funding levels, these early efforts appear modest. The annual budget for fusion averaged around $1 million until 1956, when it rose to just over $7 million (see

Appendix B). By 1958, funding had risen to $29 million. The program employed only 8 people in 1952, but then jumped to 110 in 1955 and over 200 in 1956 (Bromberg 1981).

Two fundamental beliefs seemed to drive the program during this period. One was the expectation that the development of practical fusion energy would be relatively quick and inexpensive. Many members of Congress felt that a prototype fusion reactor was only 5 or 6 years away. One official, a former director of AEC's fusion office, even predicted in 1958 that a "full-scale, power-generating" device would be operating "within a decade or two" (Bishop 1981, 170).

Secondly, after fusion research was officially declassified in 1958, and it became possible to compare the U.S. program with those of the British and the Soviets, concern arose over America's international scientific prestige. Still believing that the development of practical fusion energy was imminent, many thought that would be a severe blow to U.S. "technical leadership" for some other nation to achieve it first (Bromberg 1981, 249; U.S. OTA 1986, 44).

Within 10 years, however, both of these views had substantially changed. Ongoing research, combined with a lack of measurable progress, had made it clear that scientists had seriously underestimated the inherent technical and scientific difficulties in achieving a fusion reaction, and that useful energy from the fusion process would be beyond the capability of any nation until at least 1990. Thus, by the end of the decade, the fusion program began to appear much less attractive. Congressional disillusionment reached its peak when the House Appropriations Committee recommended a 16 percent cut in the program's FY 1963 budget, although some of this money was eventually restored.

For most of the 1960s, the MFE program more closely resembled a scientific research project than one aimed at developing a new source of energy. Work proceeded in relative quiet, emphasizing basic research in plasma behavior, as opposed to producing a prototype reactor. Even the demonstration of the tokamak by the Soviets in 1968, a momentous event in the history of fusion technology (U.S. OTA 1986, 44–46),[5] had little immediate impact on the U.S. program. The budget for fusion research throughout the decade remained stable, averaging between $20 and $30 million per year. Since the estimated cost of a tokamak at the time was around $500,000, most laboratories were reluctant to invest in the new technology. Consequently, the first U.S. tokamaks were built by the inefficient method of converting existing machines, meaning that it was some time before scientists were able to take full advantage of the new development.

The political climate for fusion changed once again in the 1970s. The onset of the energy crisis (particularly the 1973 oil embargo), the environmental movement, and the growing public unease over nuclear fission technology all served to make nuclear fusion, with its promise of abundant, clean power seem like an attractive energy option (U.S. OTA 1986, 46–50). Congressional support rose dramatically, with funding between 1973 and 1979 increasing more than ten-fold.

As MFE's political importance grew, Robert Hirsch, the project's director within AEC, began to push for another reorientation of research efforts, so that the program once again placed a greater emphasis upon actual energy production. This suggestion proved to be quite controversial. Many fusion scientists felt that Hirsch's plan, which called for pushing ahead with the engineering device before all of the physics of the process were fully understood, was a serious mistake which would cost the program in the long run.

On the other hand, this new approach was clearly in step with the times. Increased public support for MFE had been based upon the belief that it provided a potential source of energy. A few researchers even felt that adding an engineering component would give the program something of a focus. Accordingly, public statements and Congressional testimony by fusion scientists always indicated that the physics of the project was well in hand, and that beginning the technical work for a demonstration reactor was an appropriate next step (National Research Council 1978; U.S. Congress, House 1981; U.S. Congress, Senate 1980). In 1974, Congress gave approval for the construction of the TFTR, the United States' largest experimental fusion reactor, and the first device designed to utilize D-T fuel.

The program's political zenith came in 1980, with the passage of the Magnetic Fusion Energy Engineering (MFEE) Act (Public Law 96–386). MFEE called for a major commitment to magnetic fusion energy, including a doubling of its budget (which was already at historic levels) over the next 10 years. Its immediate goal was the construction by 1988–89 of a Fusion Engineering Device (FED) at Oak Ridge, designed to explore the technical problems associated with fusion-generated electricity, and to complete an assessment of its commercial potential by early in the 1990s (U.S. Department of Energy 1983; U.S. OTA 1986, 50–51; U.S. Congress, Senate 1989). The Act's long-term goal was to make possible the commercialization of fusion power possible by the first decade of the 21st century. MFEE passed with overwhelming majorities in both the House and Senate.

In spite of the Act's recommendations, Congress did not appropriate any additional funds. In fact, the fusion program's budgetary high point had come a few years earlier. Soon after management of the program had passed to the Office of Energy Research of the newly-created Department of Energy in 1977,[6] it became clear that nuclear fusion could not meet the nation's short-term energy needs. Beyond the rhetoric of the Magnetic

Fusion Energy Engineering Act, Congressional enthusiasm for the technology (as indicated by the level of funding) began to wane. Although its
budget increased in current dollars, real funding for the MFE program
began dropping in 1979. Plans for the proposed FED—the centerpiece of
MFEE—were quietly shelved in early 1981.

From the standpoint of the fusion community, the seeming inconsis-
tency of funding has had serious ramifications. During the height of the
program's popularity (and while the budget was still growing), some
fusion scientists expressed concern that virtually all major MFE research
was being carried out on the tokamak. Concern that too much emphasis
was being placed upon a single design prompted a search for alternative
techniques, thereby creating a "mixed" strategy for fusion development.
One result was the "magnetic mirror," an open tube with powerful
magnetic "plugs" at each end. These plugs "reflect" the plasma away from
the ends, thereby (in theory) reducing energy loss (Glasstone 1980, 20–
22; U.S. OTA 1986, 63–64). In 1981, Lawrence Livermore Laboratory began
to employ this technique in the construction of the Magnetic Fusion Test
Facility (designated MFTF-B). Completed in 1986, MFTF-B is the largest
non-tokamak MFE device in the world. Its contribution to MFE research,
however, has been relatively small: the budget cuts of the mid-1980s forced
Lawrence Livermore authorities to mothball it before it had ever been
used experimentally (U.S. OTA 1986, 63–65).

The funding decline continued during the Reagan Administration.
During President Reagan's first term, Congress reduced funding for virtually all non-defense R&D projects (although, as noted earlier, fusion
was considered a special case, and so was not cut as heavily as were some
projects). In addition, by the middle part of the decade, the growing
federal budget deficit had become a major national issue. Fusion research,
given its high capital costs and the fact that it was not expected to become
part of the nation's energy network for decades at least, could hardly
expect to escape unscathed. Between 1979 and 1989, its budget (in real
dollars) had fallen by 44 percent (Hamilton 1990, 501; U.S. OTA 1986, 51).

Despite the drop in political and monetary support, DOE officials
continue to promote the program. In 1986, the Department drew up the
long-range MFE strategy described in the previous section: D-T experiments leading to break-even with TFTR during the 1990s, followed by the
achievement of sustained ignition with CIT around 2000, leading to
engineering tests and development work with the ITER (U.S. Department
of Energy 1989; U.S. Congress, House 1988). Although certainly subject
to change (particularly in light of the FY 1991 budget), this stands as the
most recent DOE plan for the eventual development of nuclear fusion
energy to be officially endorsed by Congress.

Once again however, budget cuts have forced several delays. The introduction of tritium into the TFTR has deferred until at least 1993, despite the fact that it was for these experiments that the reactor was built in the first place. In addition, funding for the initial construction of the CIT has been delayed for the last three years. Although partially attributable to DOE's uneasiness over the issue of plasma transport discussed earlier (U.S. Department of Energy 1986, see note 3) much of this retrenchment has been due to Congressional concern over the federal budget deficit.

This concern was clearly in evidence in the summer of 1989 when Robert Hunter, then DOE's Director of Energy Research, submitted a plan to Congress that would once more fundamentally alter the entire fusion energy program. Hunter argued that, considering the complexity, expense, and recent technical problems of the MFE program, it was unwise to base all future reactor designs upon one particular fusion technology. He proposed that part of the funding for magnetic fusion research be diverted into civilian ICF programs, and that both approaches be pursued simultaneously as potential energy-producing technologies. The eventual goal of this redistribution would be to bring ICF up to MFE's level of development (or, put another way, to overcome MFE's 10 year head start) by the year 2000. DOE could then choose which design appeared more promising, and proceed on that basis in constructing a demonstration reactor (U.S. Congress, Senate 1980; Crawford 1989, 1434–1435). Like the ill-fated magnetic mirror, Hunter was proposing that the inertial confinement approach could serve as part of a mixed development strategy, thereby increasing the likelihood that fusion technology (in some form) would succeed.

Controversy over the Hunter plan brought the U.S. fusion energy program to a complete halt. While the magnetic fusion community (predictably) attacked the proposal as technologically unsound, members of Congress were more concerned with its impact on the budget (U.S. Congress, House 1989; U.S. Congress, Senate 1980). Adverse to funding two such programs, reluctant to embark upon yet another complicated and expensive technology, and unhappy with the program's numerous delays and setbacks, Congress called for a reevaluation of the entire fusion program. In late 1989, the Secretary of Energy appointed an outside panel of scientists and engineers to examine U.S. nuclear fusion policy and offer its "best judgement on the optimal way to structure the overall U.S. fusion program" (U.S. Department of Energy, "Charter," 1989).[7]

That panel, the Fusion Policy Advisory Committee (FPAC), completed its work in the fall of 1990. Its report essentially endorsed the development of fusion energy in general, and a mix of MFE and ICF approaches in particular. Whereas the Hunter plan called for a simple redistribution of the present fusion budget, however, the FPAC report urged that funding levels be doubled over the next 5 to 7 years. Arguing that the current spending on fusion is "inadequate even to utilize fully existing U.S. facilities and talent let alone the expanded effort needed," the panel recommended the establishment of as many as four new centers for fusion research, an ICF ignition experiment, the reinstatement of funding for the CIT, and an expansion of the U.S. role in the ITER. Total cost for this research agenda has been estimated at $3 billion. The ultimate aim of U.S. policy, according to FPAC, should be to develop an electricity-producing demonstration reactor by 2025 and a commercial fusion power plant by 2040 (Crawford 1990; Goodwin 1990, 51).

Understandably, these recommendations were warmly received by the fusion community. Secretary of Energy James Watkins, however, was less enthusiastic. In a letter to the FPAC chairman, he stated that "I am concerned that the report . . . may raise expectation in the scientific community that such a path will be easy to implement in the present budget environment . . . This is not likely to be the case" (Goodwin 1990, 51).

Watkins' prediction ultimately proved correct. In its final budget for FY 1991, Congress slashed DOE's budget for fusion by nearly $50 million—roughly 18 percent. In addition to further delays on advanced projects such as the CIT, the D-T experiments with the TFTR, and continued U.S. participation in ITER, these cuts may well force one or more fusion laboratories to close down (Hamilton 1990; Charles and Crawford 1990, 15). As a result, the future of U.S. fusion policy has been thrown very much into doubt.

The Politics of Technological Choice in Congress: A Theoretical Framework

In setting the nation's R&D priorities, Congress is making what amounts to major investment decisions: delaying spending for current budgetary needs (i.e., "consumption") with the hope of achieving greater benefits at some point in the future. For private organizations, investing in research and development is a relatively straightforward procedure. Because a firm is primarily concerned with profitability, these decisions are usually based upon such factors as risk, start-up costs, and lead times. By developing estimates of these factors, and weighing them against the projected payoff, private companies are able to make more or less "rational" investment choices. Those ventures requiring heavy initial investment or longer development periods, or which involve highly uncertain technologies cannot be readily analyzed in this fashion, and are thus less likely to attract private funding.

Clearly, the U.S. fusion energy program is the type of project that private investors would normally avoid. The exceedingly large start-up costs, the relative lack of scientific information and the longer time frame mean that fusion technologies cannot be assessed by any procedure even approximating rational choice. Despite this fact, both Congress and executive agencies have routinely made a number of crucial choices with respect to the program.

The literature in decision theory helps to provide a framework for understanding how policy makers might approach issues of this type. Theorists have long been aware that if the cause and effect premises of a given problem are unknown or in doubt, decision-makers will rely upon value premises instead. In other words, when forced to make choices in areas characterized by a high degree of factual uncertainty, both individuals and organizations tend to "fill in" the gaps in their factual knowledge with beliefs derived from some set of core values (Allison 1971; Steinbruner 1977). Empirical research has found that such value-based simplification routinely occurs whenever the facts of a given situation are unclear (Allison 1971; Coulam 1977; Kuklinski, Metlay, and Kay 1987, 615–642; Steinbruner 1977, ch. 6–10). While choices made in this fashion occasionally turn out to have been "incorrect" (when viewed in hindsight), it is frequently the only alternative to doing nothing at all.

Congress is, of course, a political body. It therefore seems reasonable to believe that the uncertainties surrounding long-term technological development will be resolved largely in accordance with prevailing political values. Moreover, since Congress is subject to a wide variety of external pressures (the electoral cycle, lobbying, constituency demands, requests from the executive branch, etc.), these internal values will likely be conditioned by forces operating within the institution's overall political, economic, and social environment.

Thus, in order to understand how Congress resolves questions of R&D policy, it is necessary to determine the value premises under which it is operating at the time. These, in turn, will generally reflect the prevailing political, economic, and social norms of society at large. In fact, the available evidence suggests that decisions with respect to developing technologies are more heavily influenced by those factors than by a project's scientific or technical status.

The U.S. space program provides the most well-documented case of this type. In 1961, Congress approved a massive project to send Americans to the moon. At the time this decision was made, space flight technology was far from proven. A number of rocket launches had ended in spectacular (and well-publicized) failure. Only one astronaut (Alan Shepard) had actually flown in space, and even that was just for 15 minutes. On the other hand, the country was still in the midst of the Cold War, and early

Soviet successes in space exploration had dealt a severe blow to American national prestige.

Ten years later, Congress sharply reduced funding for the space program. By this time, the U.S. had developed a reliable space vehicle—the Apollo/Saturn transportation system—that was still celebrating its greatest triumph. Public opinion, however, was being shaped by Vietnam, the urban riots, and a general distrust for large and expensive public ventures, particularly those directed "so far from home." In each case, Congress' decision took far more account of the dominant political mood than the technology's state of development (Hutton 1981; Logsdon 1970; Logsdon 1984).

Similar factors are clearly operating with respect to nuclear fusion. From 1951 to 1963, for example, the program's budget steadily increased, despite the fact that the scientific and technical gains during this time were relatively small. An impatient Congress did eventually cut back on funding for fusion in 1964, but 12 years of budgetary growth for a project that cannot show any progress would normally be considered highly unusual. By contrast, five years elapsed between the demonstration of the tokamak in 1968–69, a truly revolutionary development, and the provision of funds sufficient for its full incorporation into the U.S. program. In fact, in the late 1960s the fusion budget (in real terms) actually declined slightly. Spending for fusion research grew dramatically in the 1970s, even though the period was not marked by any particular scientific or technological breakthrough. When the program began to show some progress in the late 1980s, producing fusion reactions within 90 percent of the breakeven point, Congress cut its budget once again.

This same pattern can be observed with respect to specific projects. Given that fusion power is a risky and uncertain technology, proposals for a mixed development strategy (i.e., one which employs alternative approaches), would seem to have some merit, at least from an engineering standpoint. This was, in fact, the reasoning behind the initial approval of funds for the MFTF-B experiment in 1981, as well as DOE's 1989 proposal to upgrade the inertial confinement fusion program. Congressional support for these initiatives, however, could be described as erratic. The MFTF-B was first fully funded, then mothballed just before it was scheduled to go into operation. DOE's proposal, which was endorsed by an outside panel of experts—a panel appointed at Congress' direction—was effectively rejected when funding for both the ICF and MFE projects were slashed for FY 1991.

In short, the fusion energy program has two seemingly unrelated histories, one in the laboratory and one in the Congress. While obviously frustrating for scientists and engineers, this situation is consistent with the discussion of value-based decision-making presented earlier. Even after 4 decades of research, nuclear fusion remains a field characterized by a high degree of factual uncertainty, with respect to both the technology itself and the role it might play in the future of U.S. energy policy. Consequently, in deciding how best to proceed, members of Congress must rely largely upon their own value positions, which, as noted above, will be drawn primarily from political sources.

Thus, at each stage of its development, Congress has reconceptualized the fusion program to fit the dominant political climate. The 1950s were marked by a widespread belief in the desirability of high technology (especially that based upon nuclear power) and in American technological superiority generally. Congress, along with many specialists in the field, accepted that controlled nuclear fusion would become reality in a relatively short time, and that it was imperative for the U.S. scientific community to lead the way. These convictions were evidently strong enough to stand against growing evidence that no nation would develop a sustained fusion reaction anytime in the near future. The fusion R&D budget grew, sometimes substantially, every year of the decade. Early in the 1960s, however, Congressional attitudes had changed. Not only was the extreme complexity of plasma behavior, and therefore of reactor development, by now impossible to ignore, but also by 1962 the focus of "international technological competition" had shifted to space flight. Accordingly, funding for fusion was reduced, relegating the program to a lower priority in the federal science budget. In effect, Congress had redefined the project as a purely scientific enterprise.

By 1973, public opinion had shifted once again. Energy supply and environmental quality were the dominant political issues throughout the decade, and fusion, along with a number of other so-called "alternative energy sources," received massive funding increases. Although there was no evidence that fusion power could play any role in meeting the country's short-term energy needs, investment in a technology that seemed to offer an unlimited supply of electricity without harm to the environment was certainly consistent with the political norms of the time.

The most recent change in political values occurred early in President Reagan's first term. By 1982, government spending, especially the federal budget deficit, had become a major issue for both elected officials and the general public. Congress responded to this concern by subjecting all expensive budget items to increased scrutiny. Members began to focus less upon the presumed benefits of the fusion program—which, practically speaking, were decades away in any event—and more on its very large (and growing) costs. Under such conditions, the program was much more difficult to justify, making some funding cuts inevitable.

In sum, the history of fusion energy R&D can be usefully broken down into four distinct "eras," each characterized by a central political "theme":

1950s—a symbol of technological progress and American scientific preeminence; 1960s—a pure research program; 1970s—part of a comprehensive energy policy; 1980s (and continuing to present)—a budget issue. It has been these themes, much more than actual technical progress, that have determined the overall course of the program. What direction fusion energy will take in the future is, of course, impossible to predict at this point.

Conclusion

As has been stated earlier, the U.S. fusion energy program is characteristic of many projects which have already begun to make up a significant portion of Congress' R&D agenda, that is, those with long development times, high start-up costs, and which incorporate present-day scientific uncertainty. Understanding Congressional action with respect to fusion is important not only because of the costs and potential benefits of the technology itself, but also because it provides some indication of how that body is likely to deal with emerging technologies in the future.

It is clear that all elected officials will be called upon to make a growing number of critical decisions involving millions of dollars with relatively little factual information. Reforms aimed at providing policy makers with greater access to scientific expertise, such as the creation of the U.S. Congress Office of Technology Assessment or the President's Office of Science and Technology Policy, cannot by themselves remedy this situation. Almost by definition, scientific and technological (not to mention economic, sociological, and political) uncertainties are an intrinsic part of so-called "cutting edge" developments. Expert opinion can only serve to underscore the uncertain nature of such projects, that is, give non-specialists a better sense of where the controversies lie. At its best, this type of guidance can give decision-makers a clearer picture of which facts are in dispute; at worst, it becomes a competition between "my experts" and "your experts" (Collingridge 1966). While it is obviously desirable for government officials to posses as much information as possible when considering scientific issues, specialized advisory bodies can only be of marginal help in resolving questions where the basic facts themselves are in doubt.

This being the case, decisions regarding these technologies must of necessity be speculative in nature. The willingness of Congress or any elected official to gamble—and continue gambling—on such programs is largely dependent upon several elements, only one of which is their actual progress in laboratory. Budgetary considerations, the entry and exit of various issues from the political agenda, and changes in the attitudes, tastes, and perceptions of the public at large can affect Congress' views of given projects.

Moreover, these non-technical factors—their relative influence as well as their substance—are always subject to change. Given the budget process, which requires that the federal government's priorities be reviewed annually, such changes may occur frequently. Thus, as in the case of controlled nuclear fusion, a program to develop a new technology—particularly one stretching across decades—may be forced to shift direction more than once during its lifetime.

Members of Congress must appraise the future desirability of proposed R&D projects on the basis of an ever-changing set of competing values. As the nuclear fusion case demonstrates, the resulting decisions have often been a source of disappointment to the science and engineering community over the years. To the extent that these actions are reflective of more general political, social, and economic attitudes, however, such value-based decision-making is consistent with Congress' role as a representative institution. Ideally, government does not develop technology solely for technology's sake, but rather seeks to identify those innovations that potentially can serve the public interest and to develop them in a socially productive fashion. Once an emerging technology comes under congressional jurisdiction, it ceases being simply a "technical" problem and becomes, for better or worse, a political issue. As such, it is prone to the same sorts of unexpected changes and priority shifts that can affect any other item on the public agenda.

Tempting though it may be for scientists and engineers to regard Congress' actions with respect to long-term technological development as unstable, irresolute or feckless, that institution's behavior can be understood properly only when viewed in its larger political context. Indeed, it could plausibly be argued that when billions of dollars of public money is involved, political values *should* play a major part in Congressional decision-making.

On the other hand, while it may be consistent with the demands of democratic policy making, this system clearly does not serve the needs of the scientific community. The successful development of technologies like controlled nuclear fusion is a formidable task even under ideal conditions. Continuous changes in program direction or levels of funding may make it impossible. Major development projects require some measure of stability if any sort of progress is to be achieved. Moreover, ventures that are expected to stretch over decades require a steady influx of graduate students (not to mention university programs to train them), which will only occur in an environment where both political and monetary support can be assured for more than a few years at a time. The start-and-stop history of the fusion energy policy would seem to suggest that for a political institution, meeting these conditions will be exceedingly difficult.

In short, the changing nature of emerging technological systems presents policy makers with an extremely unpleasant dilemma: can democratic governments *successfully* develop highly expensive, large-scale technologies with long lead times in ways that are also consistent with their political principles? Considering not only the case of nuclear fusion, but also that of the space shuttle program and the space station, the record so far is not encouraging. As these types of projects come to make up a larger share of the U.S. R&D agenda, the abilities—both political and technical— of governmental and scientific institutions are going to be seriously challenged in the years ahead.

TABLE 6A.1   Historical Magnetic Fusion Funding, 1951–1991 (millions of current dollars)

Year	Budget Authority	Year	Budget Authority
1951–1953	1.1	1973	39.7
1954	1.8	1974	57.4
1955	6.1	1975	118.2
1956	7.4	1976	166.3
1957	11.6	TQ[a]	52.9
1958	29.2	1977	316.3
1959	28.9	1978	332.4
1960	33.7	1979	355.1
1961	30.0	1980	350.3
1962	24.8	1981	393.6
1963	25.5	1982	451.3
1964	22.6	1983	461.3
1965	23.1	1984	468.4
1966	23.1	1985	429.4
1967	23.9	1986	361.5
1968	26.6	1987	341.4
1969	29.7	1988	331.2
1970	34.3	1989	340.7
1971	32.2	1990	320.3
1972	33.3	1991	275.0

[a]Funding level for transitional quarter, reflecting change in U.S. fiscal year.

Sources: U.S. Congress, Office of Technology Assessment (1986), Starpower: The U.S. and the International Quest for Fusion Energy (Washington, D.C.: OTA), p. 210; U.S. Department of Energy, Office of Energy Research (1990), "Magnetic Fusion Energy Program" presentation to the Fusion Policy Advisory Committee.

Appendix: Abbreviations and Acronyms

AEC	Atomic Energy Commission
CIT	Compact Ignition Tokamak
D-D	deuterium-deuterium fusion reaction
DEMO	Fusion Demonstration Facility
DOE	Department of Energy

D-T	deuterium-tritium fusion reaction
FED	Fusion Engineering Device
FPAC	Fusion Policy Advisory Committee
ICF	Inertial Confinement Fusion
ITER	International Thermonuclear Engineering Reactor
JET	Joint European Torus
MFE	Magnetic Fusion Energy
MFEE	Magnetic Fusion Energy Engineering Act
MFTF-B	Magnetic Mirror Fusion Test Facility
PPPL	Princeton Plasma Physics laboratory
TFTR	Tokamak Fusion Test Reactor

Notes

This research was supported in part by Northeastern University's Research and Development fund.

1. All monetary figures are given in current dollars.

2. In a 15 June 1989 letter to 13 members of Congress, Secretary of Energy James Watkins cites the transport problem as a reason for removing CIT from the FY 1990 budget.

3. Although trace amounts are continuously deposited in the upper atmosphere by cosmic radiation bombardment.

4. Some scientists have proposed utilizing helium-3, an isotope of helium that is laid down on the surface of the Moon by solar wind, as a reactor fuel (See Horgan).

5. Indeed, Soviet results were so startling that many Western scientists speculated that they might have been fabricated.

6. After the AEC was disbanded in 1974, the MFE program was shifted to the Energy Research and Development Administration (ERDA) until the agency was abolished to make way for DOE.

7. Letter from Secretary of Energy James Watkins to H. Guyford Stever, Chair, Fusion Policy Advisory Committee, 2 March 1990.

Bibliography

Allison, Graham. 1971. *Essence of Decision*. Boston: Little, Brown and Company.

Bishop, Amasa S. 1981. *Project Sherwood: The U.S. Program in Controlled Fusion*. Reading, MA: Addison-Wesley

Bromberg, Joan Lisa, 1981, *Fusion: Science, Politics, and the Invention of a New Energy Source*. Cambridge, MA: MIT Press.

Carey, John. 15 October 1990. "Hot Fusion is Burning Dollars—and Little Else." No. 3182 of *Business Week*: 62.

Charles, Dan and Mark Crawford. 27 October 1990. "Science Survives Congressional Carving Knife." No. 1740 of *New Scientist*: 15.

Collingridge, David. 1986. *Science Speaks to Power: The Role of Experts in Policy Making*. New York: St. Martin's Press. Coulam, Robert F. 1977. *Illusions of Choice: The F-111 and the Problem of Weapons Acquisition Reform*. Princeton, NJ: Princeton University Press.

Crawford, Mark. 23 June 1989. "Fusion Plan Ignites Controversy at DOE." *Science*: (244): 1434–35.

———. 13 July 1990. "Fusion Panel Drafts a Wish List for the '90s." *Science*: (249): 119.

Glasstone, Samuel. 1980. *Fusion Energy*. Washington, D.C.: U.S. Department of Energy, Energy Information Center.

Goodwin, Irwin. September 1990. "DOE Panel Makes Case for Heating Up Fusion Research in 1990s." *Physics Today* 43(9): 51.

Hamilton, David P. 26 October 1990. "Energy Science Takes a Heavy Budget Hit." *Science* (250): 501.

Horgan, John. February 1989. "Fusion's Future." *Scientific American* 260(2): 25–28.

Hutton, Richard. 1981. *The Cosmic Chase*. New York: Mentor.

Kuklinski, James H., Daniel S. Metlay, and W.D. Kay. 1982. "Citizen Knowledge and Choices on the Complex Issue of Nuclear Energy." *American Journal of Political Science* 26(4): 615–642.

Logsdon, John M. 1970. *The Decision to Go to the Moon*. Chicago: University of Chicago Press.

———. 1984. "The Evolution of Civilian In-Space Infrastructure, i.e., 'Space Station,' Concepts in the United States." In *Civilian Space Stations and the U.S. Future in Space*. Washington D.C.: OTA.

National Research Council. 1978. *Controlled Nuclear Fusion: Current Research and Potential Progress*. Washington, D.C.: National Academy of Sciences.

———. 1989. *Pacing the Magnetic Fusion Program*. Washington, D.C.: National Academy Press.

———. Committee on Physical Sciences, Mathematics, and Resources. 1990. *Review of the Department of Energy's Inertial Confinement Fusion Program* Washington, D.C.: National Academy Press.

Niu, Keishiro. 1990. *Nuclear Fusion*. Cambridge: Cambridge University Press.

Steinbruner, John. 1977. *The Cybernetic Theory of Decision*. Princeton, NJ: Princeton University Press.

U.S. Congress, House. 29 June 1981. *Fusion Advisory Board*. Meeting convened by the Subcommittee on Energy Research and Production of the Committee on Science and Technology. 97th Cong. 2nd sess.

———. 28 October 1987. *OTA Magnetic Fusion Report "Starpower."* Hearings before the Subcommittee on Energy Research and Development and Subcommittee on International Scientific Cooperation of the Committee on Space, Science, and Technology. 100th Cong. 2nd sess.

———. 30 March 1988. *FY 1989 Department of Energy Authorization (Magnetic Fusion Energy)*. Hearings before the Subcommittee on Energy Research and Development of the Committee on Science, Space, and Technology. 100th Cong. 2nd sess.

———. 26 October 1989. *Fusion Energy Program: Status and Direction.* Hearings before the Subcommittee on Investigation and Oversight of the Committee on Science, Apace, and Technology. 101st Cong. 1st sess.

U.S. Congress. Office of Technology Assessment. 1986. *Starpower: The U.S. and the International Quest for Fusion Energy.* Washington, D.C.: OTA.

U.S. Congress. Senate. 8 July, 5 August 1980. *Magnetic Fusion Engineering Act of 1980.* Hearings before the Subcommittee on Energy Research and development of the Committee on Energy and National Resources. 96th Cong. 2nd sess.

———. 14 June 1989. *Magnetic Fusion Research and Development and Demonstration.* Hearings before the Subcommittee on Energy Research and Development of the Committee on Energy and Natural Resources. 101st Con. 1st sess.

U.S. Department of Energy. June 1983. *Comprehensive Program Management Plan (CPMP) for Magnetic Fusion Energy.* Report submitted to House Science and Technology Committee by the Secretary of Energy pursuant to the Magnetic Fusion Energy Engineering Act.

U.S. Department of Energy, Energy Research Advisory Board. November 1986. *Report of the Technical Panel on Magnetic Fusion of the Energy Research Advisory Board.* Report prepared for the Department of Energy.

U.S. Department of Energy. Office of Energy Research, Magnetic Fusion Advisory Committee, Panel 22. 1989. *Report on Confinement and Machine Physics.* Report prepared for the Office of Energy Research.

U.S. Department of Energy. Office of Energy Research. 22 March 1990. *A Plan for the Development of Magnetic Fusion Energy.* Draft report to Fusion Policy Advisory Committee.

7

The Limits of Biomedical Technology Assessment: Values, Time, and Public Expectations

Robert H. Blank

In 1976 Congress asked the Office of Technology Assessment (OTA) to examine "federal policies and existing medical practices to determine whether a reasonable amount of justification should be provided before costly new medical technologies and procedures are put into general use." Despite considerable debate over this topic since that time, technology assessment (TA) of biomedical developments remains unsatisfactory. This chapter will attempt to provide a basis for a better understanding of the reasons for this perceived inability to assess biomedical policy.

Primary attention will be directed toward: (1) the particular challenges of technology-related policies; (2) the inability or unwillingness to assess long-term problems associated with technology; (3) the political constraints on technology policy assessment; and (4) the failure to represent in any consistent manner the interests of future generations. This latter problem of representation will take the discussion into the realm of democratic theory and the extent to which a pluralist democracy can "represent" future generations.

One component of this chapter will be to analyze OTA work on biomedical policy assessment, with particular emphasis on the problems of values and time frame. I will also analyze the current status of biomedical TA activities of the Institute of Medicine, Office of Health Technology Assessment, the Prospective Payment Assessment Commission, and the estimated forty-five other organizations involved in biomedical TA of some type. This chapter will also make suggestions as to how TA might be modified to better deal with the complex policy issues raised. It will argue for the need for expanding the time frame, integrating the value dynamics and altered public expectations, coordinating the myriad of current assessment activities, and providing more critical assessments of what might be popular technologies.

Assessing Biomedical Technology:
The Value and Political Context

The challenges of technology-related policies are well illustrated in biomedicine due to the uncertainty of both future technological developments and the impact of specific attempts to influence development on a technology. The complexity of the interaction between biomedical technology and values requires considerably more attention to the long-term power of the technologies to alter values, often in unanticipated directions. Much of the current TA in this area tends to be linear in nature with little appreciation of the interrelationships and dynamics of biomedical technology, politics, and values. As a result, efforts at assessing biomedical technologies tend to underestimate technology's impact on public expectations and usage. Although it is common for many TAs today to include a chapter on ethical issues, most ignore or downplay the technology/ value dynamics and assume a static ethical framework.

The strong predisposition in our society against precluding or even slowing development of potentially life-saving technologies makes biomedicine a particularly sensitive area for critical TA. Moreover, the unrealistic dependence of the public on technological fixes to health problems, and the search for quick cures at the expense of prevention and health promotion in the U.S., explains the reluctance of Congress to deny funding even for unproven biomedical innovations. For a defense of "hightech" medicine and a reflection of this value preference see Ginsberg, who concludes that the restraint of medical technology is "not a preferred strategy" to pursue (1990, 1822). The recent reversal by Congress, for clearly political motives, of a National Institutes of Health decision to terminate funding for research on the totally implantable artificial heart, despite its dismal record, illustrates the difficulty of setting objective priorities in the highly charged area of biomedicine. Likewise, the current demand by AIDS activists (and the burgeoning AIDS research community) for quick delivery of untested drugs and for levels of funding for AIDS research that are way out of balance with funding for other diseases shows why policy makers have a difficult time setting limits on the development of biomedical technologies.

Any attempts to stop or slow development of particular biomedical innovations face vehement criticism from those individuals, interest groups, and economic interests that have a personal stake in continued funding. For instance, Fuchs and Garber (1990, 673) conclude that:

Admiration for the new technology assessment is not universal. Many practicing physicians believe it will further erode their ability to practice as they deem best; similarly, medical researchers, pharmaceutical manufacturers, and producers of medical devices fear that it will inhibit the development and diffusion of new forms of technology.

No matter how few people might benefit from a new medical procedure or technique, it is politically unattractive to recommend that it not be funded, because that decision will often result in the death of some persons either identifiable or statistical. No matter how much we might agree that aggregate health care costs must be contained, when specific proposed cuts are perceived as adversely affecting our chances of getting needed medical care for ourselves or loved ones, it is likely that we will be unsympathetic and even hostile to such efforts. Despite the general knowledge that we have finite medical resources (money, blood, organs, skilled personnel, technology), there is a presumption that each individual patient has the right to all the resources necessary to remain alive. The policy dilemma is that as long as policy makers talk in general terms about the need to cut health care expenditures there is public support. As soon as the targets of such cuts are made explicit, as they must be at some stage, policy makers face the immediate wrath of stakeholders. The likelihood of a technology assessment concluding that procedure X not be funded being found acceptable to policy makers becomes quite remote, no matter how well grounded and objective it may be.

The result of these value preferences is an almost universal failure of TA in this area to recommend against development of questionable techniques such as the artificial heart, or to reassess older technologies and consider discontinuing their use (Banta and Thacker 1990, 236). Part of this problem might stem from an inherent difficulty of TA to deal with futuristic problems. Whether because of short-term political pressures, the difficulty of forecasting long-term problems, or some combination of both, the time frame of TA continues to be limited to the near future. Moreover, the strong preference of the public and leaders for more and more advanced biomedical interventions makes any attempts to restrict their development politically unattractive. The burden these technologies may place on future generations and the negative consequences that might accompany them are thus minimized or absent from most assessments.

Technology Assessment Mechanisms and Biomedical Policy

Assessment of medical technologies has received considerable attention in the last decade, both in the public and private sectors. Seymour Perry

(1986) estimates that at least 45 organizations are involved in biomedical technology assessment, the most prominent of which are the Clinical Efficacy Assessment Project (American College of Physicians); the Diagnostic and Therapeutic Technology Assessment Program (American Medical Association): the Medical Necessity and Technology Evaluation and Coverage Programs (Blue Cross/Blue Shield); and the Hospital Technology Service Program (American Hospital Association). In addition, many medical professional societies, health provider organizations, nonprofit health-related organizations, university health institutes, and manufacturers of drugs and medical devices have formal or informal capacities for evaluation and assessment.

Despite the scope of these activities, there is little or no coordination, cooperation, or even exchange of information among the many organizations making these assessment efforts.

Although many have realized that better information on the benefits, risks, costs, and social implications of medical technology is essential to guiding the development and use of technology without unnecessarily impeding innovation, progress toward developing a coherent system for assessing medical technologies has been slow (OTA 1982, 91).

Moreover, contemporary assessments are usually narrowly focused and rarely entail the collection of primary data. They largely fail to place the specific assessments in the broader context of the national health crisis. This is not surprising, because each organization has a stake in the results of its assessment. Furthermore, none of these organizations has the resources to support necessary research, including clinical trials. Although there appears to be a realization that a more coherent system of medical technology assessment is essential if we are to make rational decisions concerning health care, genuine collaboration between these groups is problematic without a major initiative from the federal government. Given the current, and most likely long-term, emphasis on cost containment, efforts to strengthen medical technology assessment, even to achieve consensus on its basic objectives and approaches, will be difficult.

Impetus for government involvement in health technology assessment came from a 1976 OTA report which concluded that assessment should be made before costly new medical technologies and procedures were put into general use (OTA 1976). This report called for the establishment of formal mechanisms for accomplishing that task and played a part in the program of the National Institutes of Health designed to develop a consensus on technical issues. The short-lived National Center for Health Care Technology (NCHCT) was an attempt to strengthen and centralize efforts to assess health care technologies. After NCHCT's demise in 1981, the OTA reiterated the importance to the nation of a rational and systematic approach to medical technology assessment.

> The most important policy need is to bring forth a rational, systematic approach from the present multiplicity of agencies and activities to promote and coordinate medical technology assessment (OTA 1982, 18).

In 1984, Congress enacted legislation signed by President Reagan (Public Law 98–551) which revised the existing National Center for Health Services Research and broadened its mandate for assessing new technology to include, not only considerations of safety and efficacy, but, as appropriate, cost effectiveness. The new name of the center became the National Center for Health Services Research Assessment/Health Care Technology Assessment (NCHSR/HCTA). This law also established a council to advise the Secretary of DHHS and the director of the center about health care technology assessment, and it instituted a council on health care technology under the sponsorship of the National Academy of Sciences (NAS), with partial governmental support.

The latter provision implemented a recommendation by the Institute of Medicine (IOM 1983) that proposed creation of a private/public organization to assess medical technology as part of the Institute. In the legislation as passed, the council is charged with promoting technology assessment and identifying obsolete or inappropriately applied health care technologies. It has responsibility for establishing a clearinghouse for information on health care technologies and assessments as well as coordinating and commissioning assessments of specific technologies. DHHS is authorized to award the NAS up to $500,000 to cover two-thirds of the costs for planning and establishing the council. Operational funds match one dollar of federal grant money with at least twice that amount from private sources.

The initiation of the council has been delayed due to the need for sufficient long-term private support, opposition from the AMA, and some initial constitutional problems with the authorizing legislation. In addition to the problem of soliciting funds and cooperation from groups in the private sector concerned with the results of technology assessment, there is the question of the degree of cooperation of council will receive from federal agencies. Without such cooperation, the council will be unable to implement its mandate to establish a clearinghouse. Perry (1986, 242) suggests that the brief experience of the NCHCT and other organizations that have attempted to solicit such cooperation is not encouraging. Finally, there remain questions as to whether a council under the aegis of the IOM, and therefore dependent on support by private sector interested parties, could render unbiased assessments.

In 1989 Congress concluded that this joint public-private sector effort was not fulfilling the perceived needs. As a result, it renamed NCHSR/HCTA to the Agency for Health Care Policy and Research and expanded its role to provide federal leadership in medical technology assessment. One of the initial charges to the new agency will be to select a priority-setting process for selecting technologies for assessment.

Another recent privately sponsored effort to establish, in the words of Balfe et al. (1985), "a broad framework for decision making that looks beyond the immediate concerns of today to the issues that will be confronting the health sector into the next century" is the Health Policy Agenda for the American People (HPA). Initiated by the AMA in 1982, it represents the combined effort of 172 private and public organizations working together to provide a broad health care policy framework while simultaneously safeguarding the essential elements of individual decision making. Phase I of the two-phase project produced 159 principles and 41 issues covering a wide spectrum of health concerns (Boyle 1984). During phase II, which began in late 1984, policy proposals consistent with these principles were developed in response to the issues. At a press conference on February 23, 1987, HPA unveiled its report (*American Medical News* 6 March 1987, 1). Among its position statements the report concluded:

> Society must come to grips with the moral and ethical questions posed by rapid developments in health care technology—who should have access to this technology, under what circumstances should technology be applied or withdrawn, and the respective roles of providers and patients in reaching these decisions.

Throughout its deliberations, HPA placed emphasis on consensus building across the wide array of groups participating in the project. As with other such efforts in the past, it is questionable whether HPA will have any substantial policy impact.

A clear example of the difficulties inherent in establishing an unbiased national assessment program is the Prospective Payment Assessment Commission (ProPac) that was established by the legislation that imposed a prospective payment system for Medicare. ProPac was included in the legislation because of concern that the Health Care Financing Administration and DHHS, in their quest to contain costs, would not pay adequate attention to technological advances if they increased costs. In order to check this perceived bias, ProPac members are appointed by the congressional OTA, even though it also advises DHHS in the executive branch.

In congruence with congressional intent, ProPac members vowed in 1985 to take an unbiased approach toward technology assessment. They suggested that adjustments may be required in the financial incentives created by prospective payments to encourage adoption of more costly but quality-enhancing technologies. Presently, the DRG system encourages adoption of cost-reducing technologies, but discourages cost-raising ones. Unbiased assessment, however, will likely result in recommendations to use some technologies which will increase costs, and thus run counter to current efforts to cut the costs of Medicare. Because of this inherent conflict in values, it is probable that recommendations by ProPac to introduce beneficial, but costly, technologies into the DRG system will be rejected, thereby throwing this entire assessment enterprise into question.

ProPac was also charged by Congress with conducting or sponsoring original assessments of medical technology that are necessary to advise the secretary of DHHS about reimbursement rates for DRGs. Not surprisingly, little funding for that task has been forthcoming. As a result, ProPac must rely primarily on published literature and existing assessments as it develops recommendations about the incorporation of new medical technologies into the prospective payment system. Although it can emphasize those appraisals that it regards as the most appropriate and fair, the lack of resources to produce original assessments weakens its role in the policy making process. To the extent that the allocation and ultimately rationing of health care will be based on technology assessment, the lack of coherent and nonpolitical assessment mechanisms will put into question any such efforts.

In 1982, the congressional OTA concluded that emerging drugs and medical devices are adequately and appropriately identified and tested, but that emerging medical and surgical procedures are not. "The most pressing need is for some routine mechanism, e.g., the reimbursement system, to identify new procedures before they are adopted" (OTA 1982, 17). Unlike other substantive areas, the reimbursement system, rather than a regulatory agency, may be the prime candidate for assessment because coverage and payment decisions by the government have become critical factors in the diffusion of medical technologies.

Despite the fact that the OTA's critical assessment of the assessment process for medical procedures is a decade old, its observations remain relevant today. No class of medical technology is adequately evaluated on a continuing basis for either cost effectiveness or social or ethical implications. Despite efforts at TA described above, there is no single organization whose mission it is to assure that medical and surgical procedures are fully assessed before their widespread use.

Furthermore, the synthesis phase of TA continues to be weak at best. Research evidence regarding the safety, efficacy, and effectiveness of emerging technologies is seldom analyzed systematically and objectively. As evidenced by the recent expansion of coverage for heart and liver transplantations and funding of AIDS research and treatment, reimbursement and regulatory decisions continue to be under the heavy influence of the political climate and clearly reflect a value system mired in the technological imperative. The OTA's 1982 statement unfortunately describes the current context of medical TA:

> Federal agencies and private insurers and organizations set policies, guidelines, regulations, and/or make reimbursement coverage determinations, many of which profoundly affect the adoption and level of use of medical technologies. Yet, their decisions are usually based on informal, subjective, group-generated norms which tend to support the status quo (OTA 1982, 17).

Had these recommendations of the OTA (1976, 1982) been implemented in a timely fashion, the framework of TA of medical procedures and technologies today would be significantly stronger.

Interestingly, the OTA itself has been unable to break through these value constraints and critically assess emerging medical technologies. The power of the status quo, usually stated in terms of maintaining a nonpartisan stance, has led to the practice of presenting a balanced set of policy options, not policy recommendations. For instance, while the OTA report on infertility (1988) includes a detailed and very valuable description of new techniques in human reproduction and of the social, legal, and ethical implications, it steers clear of recommending any limits on the development or diffusion of these revolutionary new technologies. Although the content of the report would seem to justify the conclusion that we would be better advised to put our scarce resources in preventive approaches instead of costly individual-oriented techniques such as *in vitro* fertilization, in the name of political objectivity no such recommendation is forthcoming. Instead of stating that Congress "should," a variety of options are presented in the form of Congress "could." Similar restraints are clear in the reports on life-sustaining technologies for the elderly (1987), reproductive health hazards in the workplace (1985), the human genome projects (1988a), and in the case study on neonatal intensive care (1987). Despite OTA awareness and advocacy of the need for objective, critical, and systematic TA of medical technologies (1976, 1982), its mandate from Congress appears to limit the capacity to recommend that a particular medical strategy be blocked or even that it be given low priority due to perceived unintended or unanticipated consequences. Although I did not read completely all of the scores of case studies, staff papers, and technical memorandums produced by the Biological Applications and the Health and Life Sciences divisions of the OTA since the 1970s, none that

I read flatly concluded that a particular procedure or technique should be blocked.

Another constraint that hampers the capacity of the OTA to undertake critical prospective assessments is that it cannot handle an issue until it is politically articulated by Congress. Despite the explicit mandate in the OTA Act of 1972 that the "basic function of the Office shall be to provide early indications of the probable beneficial and adverse impacts of the applications of technology," the OTA has no autonomous power to initiate assessments. All such studies must be initiated by a congressional sponsor. The result is that studies and the reports must be in the political mainstream. In addition to undercutting the early warning function, another consequence is that the agency "tends to avoid more ambitious—and controversial—intellectual questions" (Winner 1989, 19).

In the 1985 Health Research Extension Act, Congress attempted to address these concerns by creating a Biomedical Ethics Board, composed of six senators and six representatives, with an equal number of members from each party. After more than a year of foot-dragging and in-fighting, in August, 1987, the Board finally appointed its advisory committee of experts in the fields of law, medicine, and ethics. This fourteen-member committee was mandated by the 1985 act to counsel members of Congress on the ethical issues arising in the delivery of health care and in biomedical research. However, as might be expected, the battle over selection of members focused mainly on their views on abortion and other life-related issues. The Board's proposed role in biomedical policy is off to a less than impressive start.

Most legislators are understandably hesitant to become embroiled in these issues. Especially if (as in the case of the U.S. House of Representatives) they are up for reelection at frequent intervals, they try to avoid controversial issues that might trigger single-issue group action against them. The political issues raised by biomedicine, therefore, seldom generate significant enthusiasm or activity until the situation becomes desperate enough to require action. Even then, the response takes the form of patchwork disaster control rather than long-range, definitive policy.

Barring widespread public demand for a comprehensive revision of health-care policy, Congress is unlikely to undertake it. Rather, Congress will meet the demands of powerful interest groups through piecemeal legislation, which is often inconsistent and counterproductive. While the health-care establishment cannot achieve all its goals, in part because of the lack of a consensus among its many components, it has been largely successful in keeping change to a minimum. Since the general public is silent or tolerant of the status quo, the most salient pressure on congressmen comes from groups opposed to restructuring of health care. In this

situation, the lack of vigorous, future-oriented deliberation on these issues
by public officials who desire reelection should not be surprising.

An Ethics of Responsibility to the Future

The revolution brought on by the new capacity for biomedical intervention challenges the very bases of western ethics. Traditional ethics assumes
that human nature remains more or less constant. It also presupposes that
the sphere of human action does not reach beyond the present and
immediate. As stated by Hans Jonas,

> All enjoinders and maxims of traditional ethics, materially different as they
> may be, show this confinement to the immediate setting of action . . . . The
> ethical universe is composed of contemporaries and its horizon to the future
> is confined by the foreseeable span of their lives (1984, 5).

Under these ethical frameworks, no one is held responsible for the unintended future effects of his or her well-intentioned acts. The timespan of
foresight, goal setting, and accountability is short, and proper conduct is
defined by immediate or near consequences only. The long-run of consequences beyond is left to chance or fate, not human action.

In the wake of modern technology, however, all of this has drastically
and irreparably changed. For Jonas (1984, 6), "modern technology has
introduced actions of such novel scale, objects, and consequences that the
framework of former ethics can no longer contain them." With the new-
found powers we have to reshape nature, disrupt the ecological balance,
and alter the human condition comes the corresponding ethical responsibility for the exercise of these powers. Biomedical technology is at the
center of this dramatic expansion of the range of human action and, thus,
the new responsibility toward future generations.

In response to modern technology, Jonas sets forth a theory of responsibility for both the private and public sphere. The axiom is that "responsibility is a correlate of power and must be commensurate with the latter's
scope and that of its exercise" (1984, x). Moreover, the discharge of this
responsibility requires lengthened foresight, a "scientific futurology." The
irreversible and cumulative character of technological intervention, unrec-
ognized in traditional ethics, mandates an extension of the relevant
horizon of responsibility to the indefinite future where the impact of these
interventions is likely to be most felt. Past technologies have shown the
vulnerability of nature to human intervention, but often only after damage
has already been done. The long-term implications of biomedical intervention, especially, demand knowledge commensurate with the causal
scale of action. Although there is no complete knowledge of the future

upon which to base our decisions, neither is there complete ignorance. Although lack of conclusive data may limit the value of TA, some observers have demonstrated that early assessments can help estimate the value of a technology and determine whether it merits further investigation (Phelps and Mushlin 1988). In any case, we can not opt out of our responsibility to the future by claiming ignorance.

This new ethics of responsibility also entails supplanting the traditional concept of reciprocal duty with that of nonreciprocity of duties. Conventional ideas of rights and duties, which assume that those who do not yet exist cannot make claims upon those persons who do, no longer hold in light of the powers we have to constrain the options of all posterity. Although reciprocity is the hallmark of traditional rights and duties, Jonas (1984, 39) contends that the "archetype of all responsible action," powerfully implanted in us by nature, is the nonreciprocal duty of parent to child. Although parents might expect reciprocity after the child reaches adulthood, it is not a precondition or motive for their responsibility owed the child. Just as parental responsibility is a one-way relationship to dependent progeny, so political responsibility of the present is a one-way relationship to the future, not the mutual relationship between independent adults. Unlike parental responsibility, under which childrearing has a definite substantive goal terminating in the independence of the adult child, however, political responsibility has no point of termination. No intrinsic terminus is set for political responsibility, or for mankind as a whole, by the nature of its object, because there is no such predetermined goal as in parenthood (Jonas 1984, 108).

Under the ethics of responsibility, unlike previous ethics, an agent's concrete moral responsibility at the time of action extends far beyond its proximate effects. How far it extends depends upon the nature of the object and on the extent of our power and prescience Jonas (1984, 107). As our capacity to intervene directly in the human condition expands and our power over future generations heightens, the time span of our responsibility widens appreciably. "In fact, the changed nature of human action changes the very nature of politics" (Jonas 1984, 9).

Especially in regard to biomedical intervention techniques, responsibility to future generations must become an integral aspect of any policy. The decisions made in the coming decades are likely not only to modify our conceptions of humanhood, but also to alter the characteristics of future individuals and the prospects of continued survival of the human race. Society can ill afford either to blindly ignore the opportunities presented by biomedical technologies or to actively pursue human medical intervention without including futuristic considerations. Many of the recipients of the benefits of today's research are tomorrow's citizens. Unfortunately, potential harmful effects of such research might irreversibly affect those same generations. Policy decisions made now, therefore, must consider, to the maximum extent possible, such concerns. Although it is impractical to make decisions based solely on concepts of obligation to future generations, the awareness that each innovation has broad ramifications on future alternatives ought to raise our consciousness regarding that goal. In the words of Wenk:

> People are part of the decision apparatus. Unless they are willing to trade off instant gratification for some vision of future benefits for humankind generally, and for their own progeny specifically, we will indeed be in difficulty. Unless the public embeds the future in its decision calculus, the political leadership will remain in the vise of the short run. The hazard then exists of action or inaction which could debase individual integrity or extinguish humanity altogether. Even before that may happen, the benign neglect of the future may undermine even the future capacity to decide (1981, 269).

The integration of a proper concern for the future into the policy making process necessitates substantial alterations in the way we as a society make decisions. This new responsibility casts doubt on the capacity of representative government, as now practiced, to meet these demands. Under a pluralist system, only present interests make themselves heard and felt and require consideration by Congress. Especially on issues as complex and emotionally charged as setting priorities for health care, single-issue interest groups become active, vocal, and influential. With few exceptions, their concern is with the near term or immediate, not the distant, future. Also, because public officials are held accountable to their constituencies of the present, future-oriented policy gives way to placating those persons and groups whose demands are loudest. Because the interests of the day hold sway, the future is nowhere represented.

> The nonexistent has no lobby, and the unborn are powerless. Thus accountability to them has no political reality behind it in present decision making, and when they can make their complaint, then we, the culprits, will no longer be there (Jonas 1984, 22).

Moreover, the current election system, with its dependence on political action committee contributions, clearly exacerbates the inability and unwillingness of Congress to look beyond the next election. As long as members of Congress are reelected on the basis of what they do for present-oriented interests, there remains a strong disincentive to become actively embroiled in controversial issues of the future. It is easier to avoid these "no-win" issues and ignore or obscure the long-term consequences of congressional inaction than recognize the ethics of responsibility to the future. Despite the genuine interest of a few individual members of Congress in framing a long-range approach to biomedicine, as an institution Congress is ill equipped to accept this responsibility. Workable strategies to overcome this impasse and ensure proper consideration of these policy problems, then, require consideration of creative, even revolutionary, innovations in assessment mechanisms.

In order to build a capacity for long-range future planning into the policy process and give us the ability to think systematically about the future, Lester Milbrath (1989) recommends the creation of a special governmental unit designed to provide society with a better understanding of where it is headed and what steps must be taken to get where we want to go. To this end, Milbrath proposes establishing, as part of the national government, a Council for Long-Range Societal Guidance. The special charge of the council would be to look to the long range consequences of proposed government actions and provide guidance to leaders and citizens. The council would engage in long-range forecasting and develop possible future scenarios. It would also monitor conditions and changes in society, facilitate social learning, enhance citizen dialogue and thinking about the issues, and make recommendations to public officials based upon thorough research and deep thought. Milbrath conceives of the council as composed of thirty generalists who have demonstrated a high capacity for thinking about broad social issues. These generalists, however, would be aided by two or more competing forecasting teams and adequate staffing to ensure an open flow of information and ideas. As a result of the council's efforts, the public interest would be "given a greater chance to become defined by careful, intellectual, holistic, long-term analysis, instead of by simplistic, sloganized appeals to short-term interests" (Milbrath 1989, 294).

Although the specifics of Milbrath's proposal might or might not be feasible or desirable, the concept of a new governmental mechanism to provide a future-oriented dialogue over biomedical issues is attractive. As long as the existing institutions fail to give adequate attention to the initiation and estimation stages of the policy process and are unable or unwilling even to set reasonable priorities and goals for society, resolution of these problems is impossible. I agree strongly with Milbrath that "well-deliberated long-range policies offer better solutions than hasty patchwork actions" (1989, 30). Unfortunately, fragmented, piecemeal, and simplistic attempts to deal with the complex problems concerning biomedical technologies are the norm. The resulting policies continue to fall far short of what is needed as we grapple with continually more difficult decisions.

Conclusions: Biomedical Technology Assessment

Despite the increasing magnitude and frequency of TA in biomedicine, most efforts continue to be flawed. Furthermore, the failure of such assessments has less to do with the capabilities of the assessing agents or even the strategies and methods used than it does with the value context underlying biomedicine. The constraints, then, are political and social and reflect the high personal stakes that are inherent in the life and death issues surrounding biomedical policy. The failure of most TA efforts to flatly reject certain directions of research or at least to place a low priority on them is not surprising given the potentially explosive nature of such recommendations. Moreover, in those few cases where such recommendations have been made, (e.g., artificial heart), the political response has been predictable and the TA has been ignored.

Most needed at this time is a reevaluation of these values that now constrain biomedical TA and the setting of social priorities. Again, this is not an idea new to OTA. The 1976 OTA report stated that "macroalternatives" to each technology being assessed should be defined. It is critical to consider alternative strategies to solve the same medical problem in very different ways and to consider the effect that the technology in question will have on the development and implementation of those alternatives.

> For example, in assessing a therapeutic technology, one might consider proposals for prevention of the disease in question. It would be legitimate, in this context, to ask how reasonable, feasible, or desirable these alternatives are and whether heavy investment in or implementation of the therapeutic technology would encourage, discourage, or complement their development and implementation (OTA 1976, 52).

While OTA recognizes that excessively detailed assessment of macroalternatives could lead to undesirable expansion of the scope of the TA being performed, ignoring such macroalternatives might result in obscuring the most desirable policy alternatives by subordinating the problem at hand to the particular technology. Because of the process by which technologies are now chosen for assessment by TA, and the often narrowly-focused congressional requests that limit the scope of a particular study, however, discussion of macroalternatives, particularly nontechnological ones, is frequently limited or absent in the final reports. Again, the strong value bias in favor of technological fixes, combined with the inherent dramatic nature of many technological interventions, leads to a hesitancy to reject the technological solutions and recommend against their development and proliferation in favor of alternative preventive strategies.

Bibliography

Balfe, Bruce E. et al. 1985. "A Health Policy Agenda for the American People: Phase I: The Principles." *Journal of the American Medical Association* 254(17): 2440–2448.

Banta, H. David, and Stephen B. Thacker. 1990. "The Case for Reassessment of Health Care Technology." *Journal of the American Medical Association* 264(2):235–239.

Blank, Robert H. 1988. *Rationing Medicine.* New York: Columbia University Press.

Boyle, Joseph F. 1984. "The Health Policy Agenda for the American People." *Journal of the American Medical Association* 249(15): 2073.

Fuchs, Victor R., and Alan M. Garber. 1990. "The New Technology Assessment." *The New England Journal of Medicine* 323(10):673–677.

Ginsberg, Eli. 1990. "High-Tech Medicine and Rising Health Care Costs." *Journal of the American Medical Association* 263(13):1820–1822.

Institute of Medicine. 1983. *A Consortium for Assessing Medical Technology.* Washington, D.C.: National Academy Press.

Jonas, Hans. 1984. *The Imperative of Responsibility: In Search of an Ethics for the Technological Age.* Chicago: University of Chicago Press.

Milbrath, Lester W. 1989. *Envisioning a Sustainable Society: Learning Our Way Out.* Albany: State University of New York Press.

Perry, Seymour. 1986. "Technology Assessment: Continuing Uncertainty." *New England Journal of Medicine* 314(4):240–243.

Phelps, C.E., and A.I. Mushlin. 1988. "Focusing Technology Assessment Using Medical Decision Theory." *Medical Decision Making* 8:279–289. Office of Technology Assessment. 1976. *Development of Medical Technology.* Washington, D.C.: GPO.

———. 1982. *Strategies for Medical Technology Assessment.* Washington, D.C.: GPO.

———. 1985. *Reproductive Health Hazards in the Workplace.* Washington, D.C.: GPO.

———. 1987. *Life-Sustaining Technologies and the Elderly.* Washington, D.C.: GPO.

———. 1987. *Neonatal Intensive Care for Low Birthweight Infants: Costs and Effectiveness.* Washington, D.C.: GPO.

———. 1988. *Infertility: Medical and Social Choices.* Washington, D.C.: GPO.

———. 1988a. *Mapping Our Genes: Genome Projects—How Big How Fast?* Washington, D.C.: GPO.

Winner, Langdon. 1989. "Getting The Most Out of OTA." *Technology Review* 92(8):19, 77.

8

The Role of Technology Assessment in Congressional Consideration of Biotechnology

L. Christopher Plein and David J. Webber

Introduction

Rapid advances in biotechnology have contributed to a host of complex and often controversial issues posing challenges to policy makers. Existing and potential applications of biotechnology in the fields of industry, agriculture, and medicine have generated widespread interest and concern for biotechnology in the congressional arena (Plein and Webber 1988, 1989). To understand and address the complexity of biotechnology and associated issues, policy makers rely on a wide range of information produced both within and around government. This chapter examines the importance of arguments, analysis, and information about the impacts of biotechnology provided in congressional hearings. While considering the broad range of technology information sources available to members of Congress, the emphasis of this chapter is on the congressional sources of information (i.e., the Congressional Research Service, General Accounting Office, Office of Technology Assessment, and committee hearings).

This chapter focuses on technology assessment information pertaining to four important issues associated with biotechnology: (1) the pending adoption of bovine somatotropin (bovine growth hormone) in the dairy industry, (2) the status of patent rights for genetically engineered animals, (3) the commercialization of biotechnology, and (4) the Human Genome Initiative, a series of federally funded programs designed to map the human gene. The importance of these issues is reflected in active consideration of these issues by Congress and in widespread interest in the issues as evidenced by reports in academic, popular press, and trade journal sources. As a result of this broad attention, members of Congress have a wide array of information sources from which to choose and use when considering various policy dimensions of biotechnology. Congressional activity on these four biotechnology issues differs in that the development of the issues is uneven. While a great deal of biotechnology policy knowledge (Webber 1990a) is available, it is only brought into congressional deliberations as a particular biotechnology issue ripens.

Biotechnology as a policy issue is sufficiently new that it has not generated a widely agreed on set of policy issues. Because of an 1986 administrative order of the Office of Science and Technology Policy establishing the "Coordinated Framework for Regulation of Biotechnology," the congressional role in biotechnology policy has been primarily: (1) oversight of agencies involved in the Coordinated Framework, (2) funding for biotechnology research, and (3) consideration of changes in related federal policies (e.g., the patentability of genetically altered animals) (Plein and Webber 1989).

Members of Congress view biotechnology from many perspectives with some members being primarily concerned about environmental aspects, some concerned about agricultural and economic implications, and some concerned about moral and ethical aspects of this new technology (Plein and Webber 1988, 1989). In an analysis of member speeches, committee concerns, and floor remarks, Webber (1990a, 209) found congressional interest in the following aspects of biotechnology: environmental risk of deliberate release; ability to predict consequences of release; regulatory effectiveness, adequacy and coordination; funding and performance of basic research; priorities of biotechnology applications; protection from biological warfare and terrorism; public opinion, debate, and support; preparation for emergencies; patent protection and tort liability; international competitiveness; and animal patenting. While Congress has not made public policy on each of these concerns, there is, as we shall see, a good deal of information about these concerns produced by congressional staff and agencies.

The analysis proceeds in six parts. First, the evolution of biotechnology as a policy issues is briefly reviewed. Second, sources of technological information produced by congressional support agencies (i.e., the Office of Technology Assessment, the General Accounting Office, and the Congressional Research Service) are briefly discussed. Third, other institutional sources of technological information (i.e., committee reports, caucus reports and seminars and reports sponsored by congressionally affiliated boards, commissions, and foundations) are reviewed. Fourth, other sources of technological information available to policy makers are outlined. Fifth, the use of technological information in each of the four policy issues identified above is ascertained. Sixth and finally, observations are made about the manner in which knowledge is generated, interpreted, and used when policy makers are faced with complex technological issues which span not only questions of science, but of economic, environmental, and ethical concerns as well.

Background on Biotechnology Policy Consideration

At the national level, biotechnology has emerged since the 1970s as a visible, at times highly salient, policy concern. It consists of numerous facets spanning ethical, economic, and environmental concerns. And, it has yet to be institutionalized in the policy making process. No committee in the House or the Senate has primary jurisdiction over biotechnology. Nor does a government agency have primary responsibility to administer policies relating to biotechnology (Plein and Webber 1989). The current biotechnology regulatory scheme, (i.e., the Coordinated Framework for the Regulation of Biotechnology), involves a variety of regulatory agencies depending on the application of biotechnology under consideration. Additionally, there has been a good deal of policy activity at the state level dealing with the development and use of biotechnology applications (Webber 1990b). In short, biotechnology is an issue without a policy niche.

Environmental, ethical, and economic issues associated with biotechnology have achieved high visibility in the policy arena over the years (Plein, forthcoming). In the mid 1970s, in the wake of calls within the scientific community to assess the potential environmental dangers of laboratory research involving genetically manipulated organisms, debate turned on the single dimension of how various types of biotechnological research could be classified in terms of safety. Since that time, issues associated with biotechnology have seen a number of products approved or undergo current regulatory review. In response to this, significant debate at the local, state, and federal level regarding the potential environmental, ethical, and economic impacts of biotechnology has ensued. During the course of biotechnology's evolution as a policy issue, eleven federal agencies have claimed some jurisdiction over the issue—at least 20 federal statutes, regulations, and guidelines have been identified as relating to biotechnology, and no less than 16 congressional committees have given the issue consideration through hearings and reports (Plein and Webber 1989, Plein 1989). In sum, the experience of policy debate surrounding issues associated with biotechnology reveals a complex and dynamic arena of interaction among policy makers and private interest (Plein and Webber 1989 and articles in Webber ed. 1988, 1990).

TABLE 8.1   Office of Technology Assessment Reports
on Biotechnology and Related Issues

New Developments in Biotechnology Series
  Volume 1: Ownership of Human Tissues and Cells (1987)
  Volume 2: Public Perceptions of Biotechnology (1987)
  Volume 3: Field-Testing Engineered Organisms (1988)
  Volume 4: U.S. Investment in Biotechnology (1988)
  Volume 5: Patenting Life (1988)

Other biotechnology-related reports
  Commercial Biotechnology: An International Perspective (1984)
  Technology, Public Policy, and the Changing Structure of American
    Agriculture (1986)
  Mapping Our Genes: Genome Projects: How Big? How Fast? (1988)
  Biology, Medicine, and the Bill of Rights (1988)
  U.S. Dairy Industry at a Crossroad: Biotechnology and Policy Choices
    (1991)

Technology Assessment of Biotechnology
in Congress Produced by
Congressional Support Agencies

There are three primary sources of technology assessment within
Congress: the Office of Technology Assessment (OTA), General Accounting Office (GAO), and the Congressional Research Service (CRS). Of these
three, the OTA and GAO conduct most of the primary investigations of
biotechnology, while the role of the CRS focuses primarily on compiling
and interpreting existing research. In the case of biotechnology, the role
of the OTA has been particularly visible with its studies garnering attention not only in the halls of Congress, but in the mass and trade media as
well.

The Office of Technology Assessment's visibility in biotechnology policy
debate stems from a series of reports it has published entitled *New
Developments in Biotechnology*. Initiated in 1987, to date five volumes
spanning such issues as public opinion, investment, patent and proprietary
rights, and environmental release have been published. Apart from the
five volume series, other OTA reports on biotechnology and related
subjects have also garnered widespread attention, especially the agency's
1984 study of international developments in the biotechnology industry.
Recently, the OTA published a study of the potential impacts of bovine
somatotropin—biotechnologically derived growth hormone—on the dairy
industry. Table 8.1 identifies relevant OTA reports.

In conducting assessment activities, OTA utilizes advisory panels comprised of public and private interests, solicits commentaries from interested parties, and employs contractors to conduct primary research. The OTA's review of biotechnology has been no exception to this practice. Some of the leading figures in biotechnology policy consideration have served as advisors, including: Sheldon Krimsky, a champion of close regulatory control over research; Joshua Lederberg, a leading researcher and promoter of biotechnology; Ralph Hardy of Du Pont; and Jerry Caulder, president of the biotechnology firm, Mycogen. Among the numerous contractors employed to conduct research for OTA biotechnology analyses have been: the polling firm of Louis Harris and Associates; Jack Doyle, a leading critic of commercial biotechnology; and the North Carolina Biotechnology Center. A wide range of interest in biotechnology policy debate have contributed commentaries in the drafting stages of OTA reports. These include: representatives of the Environmental Protection Agency, the Department of Agriculture, and the Food and Drug Administration; representatives of biotechnology and pharmaceutical firms, state legislators, trade group representatives, university researchers; and critics and opponents of biotechnology.

Inclusion of these varied perspectives reveals OTA's desire to include active participants in the policy arena in the drafting of technological assessments. It also reveals the desire to legitimize the reports in political terms and to prevent later challenges that reports are biased. For example, when the OTA's report on "Field-Testing Engineered Organisms" came under fire by the Food and Drug Administration, prior to its release, on the grounds that the report would be damaging to the biotechnology industry, the OTA responded that the report had already been reviewed by industry scientists who offered no serious objections to it (Flora 1987a, 1).

Reference to OTA biotechnology reports are frequently found in the public records of Congress (see, for example, U.S. Congress, House 1988a, 1–2; Congressional Record 1988, E 1381 and 1988, S 8064). But OTA's reports on biotechnology have also received significant coverage in the mass media and trade press. For example, OTA reports on biotechnology have been referred to in such popular sources as the *Christian Science Monitor* (Doyle 1987) and the *Atlantic* (Baskin 1990). In the biotechnology trade press, the OTA reports have been the subject of scrutiny and editorials. Assessing OTA's assessment performance, an editorial in *Bio/Technology* observed that, "Through the OTA, the U.S. Congress has mounted what may be the world's most impressive biotechnology information gathering and digesting organism," but also observed that, "Naturally, OTA's data are never wholly consonant with our view of the biotechnologies. And the recommendations to Congress almost always

TABLE 8.2   General Accounting Office Reports Relating to Biotechnology

The U.S. Department of Agriculture's Biotechnology Research Efforts (1985)

Agriculture's Regulatory System Needs Clarification (1986)

Biotechnology: Analysis of Federally Funded Research (1986)

Biotechnology: Managing the Risks of Field Testing Genetically Engineered Organisms (1988)

Biotechnology: Backlog of Patent Applications (1989)

World Agriculture: Factors Influencing Trends in World Agricultural Production and Trade (1989)

include options we regard as unpalatable, if not utterly insane" (McCormick 1988, 855). The weight that is given to OTA's assessments was reflected in articles (Flora 1987a, 1987b) appearing in *Feedstuffs*, an agribusiness newspaper, warning that a pending OTA report would have grave consequences for the biotechnology industry.

The reports of the General Accounting Office have not generated the same public visibility that the Office of Technology Assessment studies have garnered. However, GAO studies on biotechnology are frequently cited within the policy arena (see, for example, U.S. Congress, House 1986a). The GAO has provided numerous reports on agricultural biotechnology related issues and is currently in the process of investigating the FDA's review of bovine somatotropin in the wake of alleged procedural improprieties (Schneider 1990, 21). Since GAO reports are often prepared at the request of committee and subcommittee chairs and members, the topics of the report give a good idea of issues thought to be important in the policy arena. Table 8.2 identifies relevant GAO reports.

The Congressional Research Service has been involved in the study of biotechnology related matters for almost two decades. Its first report on genetic engineering was published in 1972 by the Science Policy Research Division. Since then, the CRS has prepared a number of reports on historical, legal, and policy aspects of genetic engineering and biotechnology. Many of these reports have been subsequently bound and made widely available. Because CRS reports are done specifically at the request of committees and members of Congress, these bound reports are published under the auspices of the committees. In the case of biotechnology reports, they have been most often published by the House Science, Space, and Technology Committee.

In addition to responding to individual member and committee inquiries, in recent years the CRS has embarked on preparing and making widely available general reports on popular issues (Vogler 1988, 232–233).

TABLE 8.3   Congressional Research Service Reports Relating to Biotechnology

Genetic Engineering: Evolution of a Technological Issue (1972)

Genetic Engineering: Evolution of a Technological Issue (1974)

Genetic Engineering, Human Genetic, and Cell Biology: Evolution of Technological Issues, DNA Recombinant Molecule Research (1976)

Genetic Engineering, Human Genetics, and Cell Biology: Evolution of Technological Issues, Biotechnology (1980)

Recent Advances in the Plant Sciences: Applications to Agriculture and Agricultural Products (1984)

Patenting Life: Issue Brief (1988)

Biotechnology: A Public Policy Primer (1989)

For example, in the midst of the animal patent controversy, the CRS prepared an issue brief entitled, "Patenting Life" (Taylor 1988), which outlined the history of recombinant DNA research and commercialization, discussed patent issues and relevant legislative action, and provided a list of suggested readings. A more ambitious project was "Biotechnology: A Public Policy Primer" (1989) which summarized the complex regulatory structure relating to biotechnology. Table 8.3 provides examples of Congressional Research Service reports on biotechnology and associated issues.

Institutional Sources of Technology Information

Apart from congressional support agencies, there are numerous institutional sources of technology assessment information within Congress. These include: congressional committee staff reports which often include detailed technical information as well as policy recommendations; reports prepared and testimony given by commissions, boards, and organizations affiliated with Congress; and congressional caucuses. Congressional staff reports are an important source of technology information for policy makers. In the case of biotechnology, a number of reports outlining the scientific, economic, and legal aspects of this new science have been produced. The advantage of committee reports rests in their ability to summarize pertinent background and immediate technical, economic, and legal information for pending legislation. For example, in August of 1988, the House Judiciary Committee submitted a report to Congress on the proposed Transgenic Animal Patent Reform Act. The report included a technical overview of biotechnological techniques, chapters of analyses by the Office of Technology Assessment and the Congressional Research Service, and summaries of hearing data and relevant legal background materials.

Another important source of information are boards, commissions, and foundations which have some form of affiliation with Congress. For example, when hearings were held on bovine somatotropin (BST) in 1986, a representative from the National Agricultural Research and Extension Users Advisory Board was asked to testify on the potential economic impact of BST. In 1989, the Administrative Conference of the United States provided a report to Congress on "Biotechnology and the Design of Regulation" which covered aspects of risk assessment and regulatory jurisdiction (see Shapiro, 1989).

In recent years, congressional caucuses have emerged as an important source of policy issue information, especially in the House (Hammond 1988). In some cases, these caucuses have become institutionalized with staff and publications (Vogler 1988, 239). One caucus, the Congressional Clearinghouse on the Future, has been visibly involved in the biotechnology issue. In 1988, the Congressional Clearinghouse on the Future sponsored a conference on biotechnology and also printed materials on the subject. In a House hearing held shortly afterwards, the conference was referred to in positive terms as a forum providing panels on emerging issues and opportunities for leaders in the biotechnology field to express viewpoints on federal policy (U.S. Congress 1988a, 1). The caucus, one of the five largest in Congress, works with a private institute—the Congressional Institute for the Future—which, in addition to sponsoring the meetings and publications, has worked with the Office of Technology Assessment in conducting technological policy assessment (Willard and Fields 1989, 23). There are other caucuses in Congress such as the Congressional Agriculture Forum, Congressional Caucus for Science and Technology, Task Force on Industrial Innovation and Productivity, and the Environmental and Energy Study Conference (Davidson and Olezek 1985, 364–365) which have a stake in biotechnology related matters, but their activities have not been as visible as that of the Congressional Clearinghouse on the Future.

Other Sources of Technology Assessment Information

Apart from institutional sources of technology information within Congress, a vast array of information sources are available to the policy maker. These include: verbal and written testimony offered by hearings witnesses, journalistic sources, and assessments offered by executive agencies. Senators and Representatives have frequently relied on newspaper and magazine articles a source of policy issue information in regard to biotechnology. For example, speaking during a hearing on BST held in 1986, Representative Tony Coelho (D-California), who initiated the proceedings, explained that the genesis of subcommittee interest in the issue stemmed from a letter and a news story that he circulated among colleagues. Coelho also commented that an editorial appearing in a major dairy trade publication, *Hoard's Dairyman*, would serve as a guide to his line of questioning in the hearing (U.S. Congress 1986b, 1–2). In the case of congressional consideration of biotechnology, news articles have been frequently reproduced in the *Congressional Record* and in hearings transcripts (see, for example, U.S. Congress, Senate 1976, 9; *Congressional Record* 1987, S 18399; U.S. Congress, House 1987, 899–890).

Commentaries by administrative agencies are often solicited by the General Accounting Office and Office of Technology Assessment in their investigations. When considering bills, committees often request that relevant agencies submit reports on proposed legislation and testify at hearings. Executive agencies also publish reports and summaries which on occasion receive wide attention. For example, in 1987, the Department of Agriculture's Economic Research Service published "BST and the Dairy Industry: A National, Regional, and Farm-level Analysis." Finding that the new animal drug would contribute to but not drastically alter changes towards greater centralization in the dairy industry, the report garnered significant attention.

Academicians also provide information about biotechnology that could be useful to policy makers in making biotechnology policy. Woodman, Shelley, and Reichel (1989) present a 350 page guide to *Biotechnology and the Research Enterprise* literature consisting of hundreds of citations. Woodman et al. organized their review according to the following eight major topics: federal research policy, university-industry relationships, conflicts of interest, university research, the biotechnology industry, international biotechnology research, related issues (public perceptions, ethical implications, economic impacts, and historical perspectives), and a directory of biotechnology organizations. Besides research and published findings, academicians also play an active role in biotechnology assessment as consultants to OTA reports and as witnesses called to testify before congressional committees.

In congressional consideration of biotechnology, private interests have played a significant role in providing technological information. As will be discussed later, the prominence of this information has varied among the issues associated with biotechnology. For example, in the case of BST, information from private sources predominated in consideration of the issue. On the other hand, in the case of the commercialization of biotechnology, the research of the Office of Technology Assessment played a key role in shaping the course of debate and policy consideration.

Biotechnology Policy Debates

The fray of debate over biotechnology moves beyond questions of technology and addresses ethical and political concerns and issues such as what impact a new biotechnological application will have on crop, meat, or dairy yields or what is the risk of a genetically engineered microbe migrating from its zone of application. At times, policy participants express issues in terms which defy compromise or reason thus limiting the use of technical information. In such a controversy laden arena, technological assessment is but one factor in a complex equation of issue consideration. In this section, we investigate four issues associated with biotechnology which have generated substantial technological assessment amidst ethical and political debate. We find that the use of institutional sources of technology information varies among the cases discussed.

The BST Controversy

Any interested policy maker will find no shortage of information on bovine somatotropin or bovine growth hormone in academic journals, government reports, and media archives. BST has been the topic of controversy since announcements were made regarding its planned introduction in the marketplace in early 1985. Designed to enhance milk production in dairy cows, the potential implications of increased production in an already over supplied market, consumer concerns over milk safety, and concerns over animal welfare regarding the use of BST, aroused debate. Also associated with these concerns was a growing skepticism of the federal government's ability to objectively regulate and review the product. Claims were made that the Food and Drug Administration and the Department of Agriculture were favorably disposed towards the interests of the biotechnology industry (Schneider 1990, A21). Table 8.4 reviews congressional consideration of the BST issue. The history of policy debate over BST is characterized by two periods of intense interest in this biotechnological product. The first period of debate occurred shortly after plans to market the product were announced. Much of the debate focused on the potential impact that the adoption and use of BST would have on the highly regulated dairy industry. In June of 1986, the House Agriculture Committee's Subcommittee on Livestock, Dairy, and Poultry held hearings on BST. The issue also gained attention in the public arena, garnering reports in such sources as the *Washington Post* (Sinclair 1986). The second period of intense interest in BST emerged in the fall of 1989 and winter of 1990. This time, much debate focused on the adequacy of the Food and Drug Administration's regulatory review of BST. While a number of

TABLE 8.4   Chronology of Congressional Consideration
of Bovine Somatotropin (BST)

1985	In October, Representative Tony Coelho (D-CA) circulates letter and an article on BST to colleagues on the House Agriculture subcommittee on Livestock, Dairy, and Poultry.
1986	In June, American Cynamid Co., one of the developers of BST, flies House Agriculture Committee Chairman E (Kika) de la Garza (D-TX) and other legislators to a briefing on BST held in Pennsylvania (Sinclair 1986, 23). Also in June, the House Agriculture Subcommittee on Livestock, Dairy, and Poultry holds hearings on the status and potential impact of BST.
1987	The USDA's Economic Research Service publishes a widely circulated report entitled *BST and the Dairy Industry: A National, Regional, and Farm-level Analysis.* The primary finding of the report is that adoption of BST will contribute to sectoral changes already underway in the dairy industry.
1988	In June, discussions regarding the impact of BST on the dairy industry figure prominently in hearings regarding the dairy sector held in Burlington, Vermont, by the House Agriculture subcommittee on Livestock, Dairy, and Poultry.
1989	Increasingly, legislation is introduced in state legislatures in order to block the use of BST.
	In July, Senator Patrick Leahy (D-VT), chairman of the Senate Agriculture Committee, calls for further tests on BST before it can be approved for the market (Larrabee 1989, 3A). Staff members of the Senate Agricultural Committee start to investigate allegations that FDA has been biased in its review of BST (Schneider 1990, A 21). The Vermont state legislature passes a joint resolution calling for Congress to study the economic effects of BST (Dairy Herd Management 1989, 6).
1990	After widely publicized allegations by a former FDA investigator that the agency had improper ties to firms developing BST, Senator Patrick Leahy (D-VT) requests that the General Accounting Office conduct an investigation of FDA's review process in January (Schneider 1990, A 21).
	On February 7, Representative Peter Smith (R-VT) introduces H.R. 4004 calling for a three-year ban on the sale of milk produced by BST-treated cows. The bill was referred to the Committees on Agriculture and Energy and Commerce.

prominent members of Congress, such as Senate Agriculture Committee Chairman Patrick Leahy (D-Vermont) publicly stated concerns over FDA's review process (Schneider 1990, 21), much of the policy action shifted to Wisconsin, Minnesota, New York, and Vermont, states where legislatures considered bills designed to thwart the introduction of BST on economic and safety grounds (Webber 1990b).

Executive agencies and private interests have played the predominate role in providing members of Congress with technological information regarding BST. Hearings held on BST in July of 1986 illustrate the dependence on executive and private sources of information. For example, a large portion of the hearing was devoted to technical aspects of BST's effect on dairy cow milk production, its impact on bovine physiology, and its effect on human safety. At the request of the subcommittee, technical information was provided in verbal and written statements by FDA and USDA officials as well as university researchers and representatives of corporations involved in the development of BST. In his testimony, FDA Commissioner Frank Young submitted over 100 scientific abstracts on biotechnology and BST to the committee and gave special emphasis to a report (Bauman et al. 1985) on BST prepared by researchers from Cornell University and the Monsanto Company (U.S. Congress, House 1986b, 40). The report was reproduced in the hearing transcript. Technical information was also supplied from representatives of biotechnology firms and university researchers. The work of at least one of the research scientists testifying at the hearing, William Chalupa, was supported by a developer of BST—the American Cynamid Company (U.S. Congress, House 1986b, 154–155).

Another critical aspect of BST has been its potential economic impact on the dairy sector. During a 1986 hearing on BST, university researchers, industry representatives, and representatives of biotechnology opposition groups provided economic analysis of the potential impacts of BST. But these issues were considered in a highly politicized context. Stressing his family's background in dairy farming, subcommittee Chairman Tony Coelho (D-California) pointedly questioned a representative of Monsanto regarding the cost of the new product (U.S. Congress, House 1986b, 107, 126). Representative James Jeffords (R-Vermont) expressed concern over contradictions between USDA's dairy herd buyout program designed to reduce milk surpluses and its support for production enhancing technologies such as BST (U.S. Congress, House 1986b, 158). Representative Steve Gunderson (R-Wisconsin) questioned the wisdom of federal research dollars supporting the development of a technology that would add to the surplus milk problem (U.S. Congress 1986b, 4).

At the time of the 1986 hearings, little congressional research had been done into the BST controversy. Primary sources of technological information were provided in large part by private actors. Public actors providing information consisted primarily of representatives of the Food and Drug Administration and the Department of Agriculture. Both of these federal bodies have been criticized for being biased towards the biotechnology industry. Only as the debate developed over BST in the late 1980s and early 1990s did both the OTA and the GAO become mobilized in conducting research on the issue. Yet at the same time, policy action on BST had shifted primarily to the states. By this time, the GAO's focus was associated with widespread concerns over the ability of the Food and Drug Administration to conduct competent and objective reviews of a broad range of drugs and food additives (Schneider 1990, A 21). This controversy led to the resignation of the FDA Commissioner Frank Young. The OTA's focus on the issue will likely consider a broader range of issues relating to BST, including animal safety, economic impacts, regulatory review, and consumer safety.

Arguably, the most salient dimension of the BST issue is the potential impact of adoption of the product on the structure of the dairy industry. There are technical dimensions to this debate, in terms of per cow production increases, regional adoption rate disparities, and economy of scale questions for individual producers. But, much of the debate has focused on the survival of the family farm in the face of modern corporate agriculture and the wisdom of introducing a production enhancing technology into an already surplus burdened market. These issues represent the nexus of the technological and political elements of policy debate. The delayed participation by such congressional agencies as the OTA in the study of the policy dimensions of BST may reflect a reluctance to become involved in the fray of debate. Historically criticized as having a liberal bias (Reppert 1988), the OTA has earned prominence in providing information regarding biotechnology in general, and has even won grudging respect from those who differ with its findings (McCormick 1988, 855). The BST debate, which is the tip of the iceberg of a much larger and enduring debate centering on the role of the federal government in agriculture, may have proven too risky an issue in which to become involved. If this is the case, then OTA's recent initiation of a report on BST may be seen as a move made possible only after a successful track record in biotechnology assessment activities had been established.

The Patent Issue

While the role of institutional sources of information was limited in the early years of the debate over BST, these sources have played a much more visible role in the debate over the patenting of genetically engineered life forms. Throughout the development of the patent issue debate, congressional support agencies have been involved with assessment activities. The most visible documents relating to this issue have been the OTA's reports "Ownership of Human Cells and Tissues" published in 1987, and "Patenting Life" published in 1989.

Two dimensions of the debate surround the patent issue. On one hand, attention has centered on legal and procedural questions that focus on the ability of the Patent Office to conduct reviews—whether biotechnological processes as well as products should be patentable—and the relationship of patenting to foreign competition. This topic has been the subject of a 1988 House hearing, a 1989 GAO report, and recent legislation introduced in both the House and Senate. The other dimension—the patenting of genetically engineered organisms—has been much more controversial and visible. Our discussion focuses on this dimension of the patent debate.

In 1980, the Supreme Court decided in *Diamond v. Chakrabarty* that genetically engineered organisms were patentable. In rendering its majority opinion, the Supreme Court emphasized Congress' responsibility to deal with patent issues and address issues raised by new technologies (447 U.S. 318, 1980). In the immediate wake of the Court decision, no action was taken in Congress to amend the patent code in regard to genetically engineered organisms. Outside of Congress however, actions by the Court of Patent Appeals and Patent Office broadened the scope of patentability to include vertebrates and mammals. These decisions, applauded by the biotechnology industry, sparked opposition from a number of interests concerned about economic, animal welfare, and ethical aspects of animal patenting. The decisions also generated significant media coverage in such sources as *Time, New Republic, Washington Post,* and *New York Times.* The Patent Office policies also sparked congressional responses in the form of bills and hearings aimed at patent reform. Table 8.5 illustrates significant events in the evolution of the patent issue.

Before the furor over the granting of patents to genetically engineered organisms erupted in Congress, the OTA had already published two reports containing information relating to the patent issue. The OTA's 1987 report, "Ownership of Human Tissues and Cells," covered the broad ethical, economic, and legal aspects of proprietary rights over biological matter. While not directly addressing issues of transgenic animal patents, the report did sketch out the context of which patent issues would likely be debated.

Of perhaps more immediate use to policy makers in the patent controversy was the OTA's 1984 report, "Commercial Biotechnology: An International Analysis." This report identified intellectual property law as a factor of "moderate importance" to the development of a healthy American biotechnology industry (OTA 1984, 16). The report dedicated an entire chapter to patent issues and was later placed in the transcripts of the 1987

TABLE 8.5   Chronology of Congressional Consideration of Patents

1980	In *Diamond v. Chakrabarty* the Supreme Court rules that genetically engineered microorganisms are eligible for patents.
1987	In March, the OTA publishes *New Developments in Biotechnology: Ownership of Human Cells and Tissues*. The Patent Office extends patent eligibility to higher forms of life in April. In June a House Judiciary subcommittee begins hearings on patents and the Constitution; additional hearings were held in July, August, and November. In August, Representative Charles Rose (D-NC) introduces H.R. 3119 calling for a two-year moratorium on animal patents, but it does not pass in Congress (Taylor 1988).
1988	In February, Senator Mark Hatfield (R-OR) introduces S. 2111 calling for a moratorium on animal patents. In June, Representative Robert Kastenmeier (D-WI) introduces two bills, H.R. 4970 and 4971, designed to grant royalty exemptions to farmers and researchers. S. 2111 passed in Senate but was not voted on in the House. H.R. 4970 passed unanimously in House but not voted on in the Senate.
	In March, the House Judiciary subcommittee on Courts, Civil Liberties, and the Administration of Justice rejects a report prepared by its own staff. The report favored current practices of allowing the federal government to allow patenting of animals (Schneider 1988, 7). Also in March, a subcommittee of the House Committee on Small Business holds hearings on the backlog of patent application in the Patent Office. The backlog of biotechnology patents is the primary focus of the hearings.
	In April, the Patent Office grants a patent to Harvard researchers who have developed a transgenic mouse. "The Harvard Mouse" decision receives wide media coverage.
	In July, the CRS publishes a report entitled "Patenting Life." Also in July, Representative Robert Kastenmeier emphasizes the importance of patent regulation in testimony before subcommittees of the House Science, Space, and Technology Committee holding hearings on the Biotechnology Competitiveness Act.
1989	In March, Representative Robert Kastenmeier introduces two bills, H.R. 1556 and H.R. 1557. The former seeks to clarify patenting procedures for animals and is referred to the House Committee on the Judiciary. The latter—The Transgenic Animal Patent Reform Act designed to regulate the use of genetically engineered organisms in agriculture—is referred to two House committees.
	In April, the OTA publishes *New Developments in Biotechnology: Patenting Life*. Also in April, the GAO delivers a report *Biotechnology: Backlog of Patent Applications* to the House Science, Space, and Technology Committee.

(continues)

TABLE 8.5 (*continued*)

In September, a subcommittee of the House Judiciary Committee holds hearings on transgenic animal patents. Also in September, Representative Benjamin Cardin (D-MD) introduces H.R. 3247 calling for a two-year moratorium animal patents.

1990   In January, Senator Mark Hatfield (R-OR) introduces S. 2169, a bill calling for a five-year moratorium on genetically altered animal patents in order to clarify economic, environmental, and ethical issues raised by the issue. The bill is referred to the Committee on the Judiciary.

In February, Representative Rick Boucher (D-VA) and Carlos Moorhead (R-CA) introduce H.R. 3957, The Biotechnology Patent Protection Act designed to eliminate "unfair" advantages held by foreign competitors. In March, Senator Dennis DeConcini (D-AZ) introduces S. 2326, a companion bill to H.R. 3957.

hearings on transgenic animals. "Commercial Biotechnology's" larger discussion of the commercial viability of biotechnology and how it might give the United States a competitive edge in the global marketplace also figured in the patent debate. In the wake of the Patent Office decisions regarding the patentability of genetically engineered animals, a number of policy makers and private interests called for a moratorium on patenting until further research into the ethical, economic, and animal welfare questions could be undertaken and completed. In 1988, legislation was introduced in both the House and Senate calling for such a moratorium. A broad coalition of interests representing animal welfare groups, environmentalists, small-scale farmers, and religious groups favored the moratorium, but the initiative failed (Plein, forthcoming). One of the coalition's leading spokesmen, Jack Doyle, Director of the Environmental Policy Institute testified in hearings that one of the reasons for supporting the moratorium was to give the OTA more time to conduct a thorough investigation of the animal patent issue (U.S. Congress 1987, 97). In an interview conducted in January of 1990, Doyle commented that the subsequent OTA report, "Patenting Life" (1989) "wasn't very good and at times was simply wrong."

Representative Robert Kastenmeier (D-Wisconsin), who initiated the hearings on patents and transgenic animals in both 1987 and 1989, commended the report. On the day the report was released, Kastenmeier introduced legislation aimed at providing patent royalty exemptions for farmers and urged colleagues to review the report. He referred to the report as "comprehensive" and identified it as addressing issues similar to those raised in his proposed legislation (*Congressional Record* 22 March

1989, H 831). In September of 1989, an official of the OTA testified before Kastenmeier's subcommittee at hearings on transgenic animals.

As the patent issue has evolved from the initial general concerns raised by *Diamond v. Chakrabarty* to more specific questions of ethics, animal welfare, economic impact, proprietary rights, and the ability of the federal government to address the issue, Congress' institutional sources of technological information have responded to these issues on relatively short order. All three institutions have been involved with analysis of the various dimensions of the patent issue. The GAO's activities have focused on the performance of the Patent Office dealing with biotechnological patent issues. The OTA's reports have reviewed the broad range of ethical and economic factors involved. The CRS has provided updated reports on patent issues and recently covered the issue in its 1989 report, "Biotechnology: A Public Policy Primer."

In many ways, the response of congressional research agencies to the patent issue has been very much like these agencies' experience with environmental release issues in biotechnology. In the case of biotechnology and patenting, each new advance or policy decision calls for assessment, which, by and large, congressional sources have provided. In similar fashion, as new techniques and products were developed for the market, adding new wrinkles to the debate over the safety of releasing genetically engineered organisms in the environment, the OTA, CRS, and GAO have played a visible assessment role. Whereas the debate over BST has been product orientated, mobilizing dedicated participants in the process, and giving the issue a clear basis of distinction—debate over patents, and for that matter, environmental release, is process oriented. In other words, the BST debate has a "win or lose" dimension to it which makes it highly controversial. On the other hand, the patent issue is a much more diffuse topic of debate which has sustained wavering levels of participation from a broad spectrum of policy participants. In such a process-oriented arena of policy consideration, non-partisan technological assessments provided by congressional research agencies would likely be more welcome than those offered in a highly controversial debate surrounding the introduction of a product like BST. For in such a climate, such reports would be likely held in suspicion by highly mobilized policy participants who already hold steadfast assessments of the issue.

The Commercialization of Biotechnology

Whereas the performance of Congress' institutional sources of technology information reflected long-term delay in reacting to issues associated with BST and systematic response to the development of patenting issues, in the case of the commercialization of biotechnology, one congressional source of technology information, the OTA, essentially placed the issue of biotechnology commercialization on the agenda and drove the course of policy consideration. Its report, "Commercial Biotechnology: An International Analysis," published in 1984, has been frequently cited in congressional proceedings, the trade press, and the mass media. A subsequent report, "U.S. Investments in Biotechnology," published in 1988, also achieved a high level of visibility in the biotechnology policy community. Table 8.6 illustrates congressional consideration of the commercialization of biotechnology.

Biotechnology has long and widely been touted as a means of promoting economic development through high-technology investment and commercial development. Much of the rhetoric surrounding biotechnology focuses on America's international lead in the field and how this lead must be maintained in the face of mounting foreign competition (Freudenheim 1988; Leahy 1988; USDA 1988). Because of the labyrinth of regulatory procedures surrounding research, development,and patenting of biotechnological products, numerous industry sources and representatives have argued for government efforts to coordinate and streamline the bureaucratic process in order to enhance the development of a commercial biotechnology industry in the United States. As a result, a number of legislative efforts have been initiated to enhance the development of the biotechnology industry. These efforts, though, have met with failure, due in large part to conflicting visions of how to best achieve the goal of American leadership in the global biotechnology market.

According to a 1986 report prepared by the House Committee on Science and Technology's Subcommittee on Investigations and Oversight, the OTA's 1984 report "Commerical Biotechnology: An International Analysis" served as an impetus for efforts in Congress to coordinate biotechnology research and regulatory activity in order to maintain American competitiveness abroad (1986a, 6). The 1984 OTA report identified a number of broad policy options that Congress could act on to enhance biotechnology industry development in specific and high-technology development in general. Perhaps most importantly, the report identified 10 factors that influenced competitiveness in the biotechnology industry spanning such issues as: federal funding of research, financing and tax incentives, personnel training, effective regulation, patent law, university-industry relationships, anti-trust law, international technology transfer, investment and trade, and coordinating the biotechnology industry through government "targeting" policies (OTA 1984, 12–20). These factors were in turn seized upon by both policy makers and participants in the policy arena.

While ranked lowest on the OTA's list of factors influencing success in the commercialization of biotechnology, the idea of targeted policies

TABLE 8.6 Chronology of Congressional Consideration of Biotechnology Commercialization

1984 Office of Technology Assessment publishes *Commercial Biotechnology: An International Analysis.*

The White House Office of Science and Technology Policy announces intentions to develop a coordinated framework for biotechnology regulation and invites public and agency comment in the December 31 issue of the *Federal Register.*

1986 Coordinated Framework for the Regulation of Biotechnology is adopted in June.

General Accounting Office publishes *Biotechnology: Analysis of Federally Funded Research for the House Energy and Commerce Subcommittee on Oversight and Investigations* in August, which outlines widespread differences in agency approaches to biotechnology research.

1987 Senate Bill 1966, The Biotechnology Competitiveness Act of 1987, is introduced on December 15. Among the bill's sponsors are Senators Lawton Chiles (D-FL), Edward Kennedy (D-MA), and Patrick Leahy (D-VT).

1988 A companion bill to S. 1966 is introduced in the House by Representative James Scheuer (D-NY) on April 29. The bill, H.R. 4502, is titled the Biotechnology Science Coordination and Competitiveness Act.

In May, the Senate Labor and Human Resources Committee reports favorably on S. 1966. The bill is approved in the Senate by an 88 to 1 vote on June 17. In the House, the bill is referred to the House Agriculture, Science, Space, and Technology, and Energy and Commerce committees.

The Office of Technology Assessment publishes *U.S. Investment in Biotechnology* in July.

In July, the Congressional Clearinghouse on the Future sponsors a conference on biotechnology that features representatives from industry, universities, and public interest groups. Several members of Congress attend, including Senator Lawton Chiles (D-FL), Senator Tom Harkin (D-IA), and Representative James Scheuer (D-NY).

In July, subcommittees of the House Committee on Science, Space, and Technology hold hearings on S. 1966 and H.R. 4502. In August, the Subcommittee on Department Operations, Research, and Foreign Agriculture of the House Agriculture Committee holds hearings on S. 1966. The Energy and Commerce Committee fails to take up the bill for consideration. S. 1966 dies at the end of the 100th Congress.

proved to be a popular issue among policy makers. In 1987 and 1988, efforts were made in both the House and the Senate to pass legislation calling for the establishment of a Biotechnology Competitiveness Act. Among the major reasons given for the need for the Competitiveness Act was the threat of foreign competition and the loss of America's lead in the international "biotechnology race." In comments made on the floor of the Senate, Senator Lawton Chiles (D-Florida), sponsor of the Biotechnology Competitiveness Act, stressed the need to develop a strategy for coordinating biotechnology research and development in the face of international competition. He also inserted into the record two articles from the *Wall Street Journal* reporting on Japan's efforts to achieve international superiority in biotechnology (*Congressional Record* 1987, S 18399). Senator Patrick Leahy (D-Vermont), a cosponsor of the act, commented on the Senate floor that, "Congress must act now to ensure American leadership in this emerging technology" (*Congressional Record* 1987, S 18403). He later commented that, "The biotech race is a race America can't afford to lose" (Kuntz 1988, 2428). Senator Edward Kennedy (D-Massachusetts), also a cosponsor of the act, referred to the *Wall Street Journal* articles as "evidence of the need to focus the attention and energy of this Nation on biotechnology" (*Congressional Record,* 1987, S 18404).

In extended remarks published in the *Congressional Record* (1988, E 1381) after introducing the House version of the Biotechnology Competitiveness Act, Representative James Scheuer (D-New York) commented that "Congress' failure to address these larger issues—to decide what our national strategy should be on biotechnology—is an abdication of our responsibility which could send the U.S. biotechnology industry to the high-technology graveyard" (*Congressional Record* 1988, E 1381–1382). Representative Ron Wyden (D-Oregon) commented, "We need a coordinated government policy or . . . foreign competition overtaking us is a real possibility" (Kuntz 1988, 2427).

In 1988, the OTA published "U.S. Investments in Biotechnology" which largely reinforced the findings of the 1984 report. Citing the OTA's report on American investment in biotechnology, Representative Claudine Schneider (R-Rhone Island) voiced at a House Science Subcommittee hearing on S. 1966 and H.R. 4502, her fears that current public and private research efforts might not be able to take advantage of the full potential of biotechnology and as a result could lead to the "loss of our international lead in this field" (U.S. Congress, House 1988a, 1).

The frequent reference to the OTA's 1984 and 1988 reports on the commercial dimensions of biotechnology reveals the agenda setting role that these reports had among interested parties in the policy process. These reports provided a touchstone in the policy process for which discussion could emanate and revolve around. Both of these reports reveal how assessments, developed in the name of objectivity, once released into the policy process become politicized artifacts in debate.

Human Genome Initiative

Since James Watson and Francis Crick described the structure of DNA in 1953, there has been a great deal of interest in conducting further research that would result in applying this genetic knowledge to improving humankind. Potential applications of such knowledge in medicine include better diagnostic procedures, better production of medicines based on limited enzymes and hormones, and better preventative medicine due to genetic identification of disease (Bishop and Waldholz 1990). The Human Genome Initiative is a federally funded program—three billion dollars over 15 years—intended to do just that. In the opening statement of the Congressional hearing on the initiative, Senator Ernest Hollings (D-South Carolina) stated the goals of the research program as follows:

> The goal of the Human Genome Initiative is to determine what makes us human and what makes each of us a unique individual. By deciphering the genetic code, we will be able to diagnose genetic diseases, better understand cancer and other often fatal afflictions, and create hundreds of new genetically engineered products. The technology developed for sequencing the human genome will enable geneticists to sequence the genes of other organisms, enabling scientists to develop better crops and farm animals. In addition, this technology will provide new opportunities in genetic engineering and biomedical research. There are many other benefits of the Initiative which can only now be guessed at (U.S. Congress, Senate 1989, 4).

The 1988 OTA report, "Mapping our Genes," examines the variety of human genome projects underway in the United States; ways to fund and to coordinate these large-scale scientific projects; and social, ethical, and international considerations of human genome projects. While the report is generally supportive of human genome projects, it identifies a number of concerns ranging from genetic determination and eugenics (OTA 1988, 82–84) to questions about the management of science (OTA 1988, 125–128).

"Mapping our Genes" reviews four sets of recommendations for congressional consideration dealing with appropriations to federal agencies supporting genome research, access to information, organization of genome projects, and questions for congressional oversight (OTA 1988, 15–17). The report identified the core issue as "resource allocation for research infrastructure" (OTA 1988, 10) and urged support for databases, maps and repositories for the high volume of information being generated by genome research.

TABLE 8.7   Congressional Consideration of the Human Genome Initiative

1986	In October, through meetings of an advisory committee, the National Institutes of Health begin efforts to coordinate human genome research (OTA 1988, 94). Representative Claude Pepper introduces legislation calling for the establishment of a National Center for Biotechnology Information in order to coordinate data bases of genetic information. The bill, H.R. 5271, fails to become law.
1987	Legislation similar to that of H.R. 5271 is introduced in both the House and the Senate. In December 1987, the proposed National Center for Biotechnology Information is incorporated into S. 1966, the Biotechnology Competitiveness Act.
1988	The Office of Technology Assessment publishes *Mapping Our Genes: Genome Projects: How Big, How Fast* in April. Congress sets aside $17.2 million NIH human genome research.
1989	National Center for Human Genome Research is established in the National Institutes of Health on October 1, 1989, under the directorship of Dr. James Watson. In November, hearings are held on "The Human Genome Initiative and the Future of Biotechnology" by the Science, Space, and Technology Subcommittee of the Senate Committee on Commerce, Science, and Transportation.
1990	In July 1990 the Senate Committee on Energy and Natural Resources Subcommittee on Energy Research and Development holds hearing on the Human Genome Project.

Table 8.7 outlines the congressional consideration of the Human Genome Initiative. Senate hearings in November 1989 to review the "Human Genome Initiative and the Future of Biotechnology" featured the testimony of Dr. James Watson, director of the project, and several other prominent scientists speaking about he promise of genome research. All of the scientists and Senators who submitted prepared statements were already involved on various genome initiative working groups or were already on record as supporting the project. It should be noted that this Senate hearing was an oversight hearing without specific legislation or appropriation under consideration at that time. Although at face value the scientific and medical goals of the human genome project appear to be benign and acceptable, this research initiative mobilized opposition from two very different sets of actors. Reflecting concerns born out of shrinking federal support for research, one set of actors representing members of the scientific community have criticized the project as a poor investment diverting funds from other worthy research pursuits. This settlement has

been expressed in specialized journals (Baltimore 1987; Ayala 1987), the mass media such as the *New York Times* (Angier 1990), and in congressional hearings. As noted scientist Bernard Davis commented before a Senate subcommittee hearing on the Human Genome Project in July of 1990:

> Although all the goals of the HGP [Human Genome Project], except for the complete sequencing of the human genome (now postponed), are clearly worthwhile, there is widespread concern over the present status and plans of the program. There are two reasons: its competition with other research funds at a time of financial stringency, and doubt that its scientific benefits justify its rapid expansion and its organization in the pattern of "big science" (Davis 1990).

Apart from some members of the scientific community, a coalition of interests are opposed to the Human Genome Project on ethical grounds. Known as the National Coalition on Gene Mapping, the coalition includes one of biotechnology's most visible critics, Jeremy Rifkin, as well as such religious interests as the National Council of Churches, and the Religious Action Center of Reform Judaism (Fox 1988, 643). Their concerns that knowledge of the human genome will be used for nefarious purposes (i.e., discriminatory genetic screening and eugenics) are reminiscent of concerns raised early in the biotechnology debate and detailed in such early technical reports as the CRS's "Genetic Engineering: Evolution of A Technological Issue" (1972). In short, the Genome Initiative demands assessments of perennial questions central to biotechnology debate as well as assessments of issues tied to the contemporary economic situation.

Conclusion

Designed to give comprehensive and objective analysis of complex policy issues, congressional support agencies unquestionably inhabit a politicized environment. The case of biotechnology and its various episodes of debate reveal the importance of these agencies in shaping the contours of policy consideration by the nature of their involvement—or lack of involvement. By developing policy knowledge, congressional support agencies become political actors in the policy process. Within the policy arena, there is the desire to maintain nonpartisanship in support agencies such as the OTA (Reppert 1988; Hershey 1989). The actions of support groups during the course of biotechnology debate reveals the political sophistication needed to cultivate this image and maintain the mission of credible assessment.

Numerous episodes in the biotechnology policy issue debate reveal the political weight of technology assessment. Pleased with OTA's "Investment in Biotechnology" report, the editor of *Bio/Technology* commented: "Actually the results of this unusual information exercise have been excellent, dating back to the 1984's 'Commercial Biotechnology: An International Analysis.' It's easy to lose track of that in the tide of venom that tends to greet each new [OTA] report" (McCormick 1988, 855). And it is with venom that OTA assessments have often been received. In the fall of 1987, for example, a high-level FDA official, Henry Miller, labeled the pending field testing report as "nothing short of a catastrophe" and called on the biotechnology industry to "raise the roof with the OTA" before the report was released for "you can exert much better leverage before the ax falls" (Flora 1987b, 1, 3). The OTA responded by criticizing Miller's actions as "very inappropriate" and as "violating the trust" placed in reviewers not to reveal OTA findings before reports are released. The OTA also emphasized that the views of concerned parties had been solicited at great length in drafting the report (Flora 1987a, 1).

There are indications that congressional support agencies have sought to avoid controversy not only through the inclusion of diverse parties but through the avoidance of debate. The absence of congressional support agency involvement in the controversial BST debate has perhaps contributed to policy consideration which has dragged on unresolved for half a decade. Considering the degree of involvement in other areas of biotechnology, this inaction may likely have been a matter of choice, especially by the relatively independent OTA, to steer clear of an unsettled and divisive issue. In such a volatile debate, analysis would tend to be disregarded. And considering the newness of many biotechnology related issues generated by commercial and scientific developments, a tarnished image in the BST debate may have undercut the legitimacy of assessment activities in other arenas of biotechnology debate. Now that congressional support agencies such as the OTA and GAO have established track records in biotechnology assessment activities, they are in a better position to offer analysis within the fray of BST debate.

Whereas as controversy was sought to be avoided in the case of BST, congressional support agencies, especially the OTA, have played an active role in driving the course of debate in the area of biotechnology commercialization and international competition. While the BST issue has been divisive, the commercialization of biotechnology has received ringing bipartisan support and has been championed by many established interests in Washington (see Plein, forthcoming). While there are numerous interests concerned about the commercialization of biotechnology on agriculture and on the environment, the popularity of this issue and the generally positive analysis assessments provided has won the grudging respect of biotechnology's supporters, who are relieved to see the assessments are not always the words of doomsayers.

While both the issue of BST and commercialization have had high visibility, the cases of patent issues and the human genome project offer examples of assessment offered in more subdued environs. In the case of patentability issues, we see assessment adjusting to steps in the progress of technology. And in the case of the Human Genome Project, assessment focuses not only on the appropriate research strategies, but also attempts to fathom the ethical questions that arise as new areas of knowledge are revealed. While the patent issue has had its moments of high public visibility, especially in the wake of the Harvard Mouse decision, the Human Genome Project is only now emerging as a visible concern. The project may pose special challenges for congressional support agencies because of interagency debate over research funding between the National Institutes of Health and the Department of Energy, as well as debate within Congress between champions of these agencies. These tensions, which have also involved the Department of Agriculture, were at the root of the failed legislative initiative to pass the Biotechnology Competitiveness Act in 1988 (Plein and Webber 1989, 195–196).

Congressional consideration of biotechnology is still in a formative stage. Biotechnology is a young policy issue marked by the absence of major congressional action. One policy making consequence of the 1980 Supreme Court's *Diamond v. Chakrabarty* decision, and of the executive establishment of the Coordinated Framework in 1986, is that it reduced the immediate need for congressional action, thereby allowing time for the production and accumulation of a body of information about biotechnology policy that stands ready to inform policy making. As Congress is largely a reactive policy making institution, its internal information sources respond to the potential that issues will evolve into policy issues. Among the four issues relating to biotechnology reviewed here, the controversy surrounding BST demonstrates this point. Despite a large volume of OTA, CRS, and GAO reports on various aspects of biotechnology, little congressional research had been done before 1986 when the BST issue began to ripen as a political controversy. This controversy appears to have stimulated these congressional information sources to examine this issue. As anticipatory producers of technology assessment, these congressional support agencies appear to do a good job of identifying important aspects of biotechnology policy issues that warrant study. Ultimately it is the evolution of a policy issue that shape both the production and use of technology assessment in Congress.

Bibliography

Angier, Natalie. 1990. "Vast 15-Year Effort to Decipher Genes Stirs Opposition." *New York Times* (5 June): B5, B8.

Ayala, Francisco J. 1987. "Two Frontiers of Human Biology." *Issues in Science and Technology* (Spring): 51–56.

Baltimore, David. 1987. "Genome Sequencing: A Small Science Approach." *Issues in Science and Technology* (Spring): 48–50.

Baskin, Yvonee. 1990. "Getting the Bugs Out." *The Atlantic* (June): 40–47.

Bauman, Dale E., Phillip J. Eppard, Melvin J. DeGeefer, and Gregory Lanza. 1985. "Responses of High Producing Dairy Cows to the Long-Term Treatment with Pituitary Somatotropin and Recombinant Somatotropin." *Journal of Dairy Science* 68: 1352–1362.

Bishop, Jerry E. and Michael Waldholz. 1990. *Genome: The Story of the Most Astonishing Scientific Adventure of Our Time—The Attempt to Map All the Genes in the Human Body*. New York: Simon and Schuster.

Congressional Clearinghouse on the Future, July 1988, "Biotechnology" *Emerging Issues*. Washington D.C.: Congressional Cleaning House on the Future.

Congressional Record, December 17, 1987, Biotechnology Competitiveness Act, S18399–S18405.

———. May 4, 1988, H.R. 4502 The Biotechnology Science Coordination and Competitiveness set of 1988, E. 1381–1382.

———. June 17, 1988, Biotechnology Competitiveness Act, S. 8061–8101.

———. 1989, "Genetically Altered Animals: Regulatory Reform and Patent Protection Issues." H830–H831 (March 22).

Davidson, Roger H. and Walter J. Oleszek. 1985. *Congress and Its Members*. Washington D.C.: Congressional Quarterly Press.

Diamond v. Chakrabarty. 1980. 447 U.S. 318. Davis, Bernard. 1990. Testimony given before the Senate Committee on Energy and Natural Resources, Subcommittee on Energy Research and Development. Hearing on the Human Genome Project. 101st Cong. 2nd Sess., July 11.

Doyle, Jack. 1987. "DNA—It's Changing the Whole Economy." *The Christian Science Monitor* (September 30).

Flora, Charles. 1987a. "OTA Report Said Damaging to Biotechnology." *Feedstuffs* (5 October): 1, 3.

———. 1987b. "Federal Agencies Enter Power Struggle Over New OTA Biotechnology Report." *Feedstuffs* (19 October): 1, 27.

Fox, Jeffrey. 1988. "Critics Converge on Genome Project." *Bio/Technology* (June): 643.

Freudenheim, Milt. 1988. "The Global Biotechnology Race." *New York Times* (13 July): 34.

Hammond, Susan Webb. 1988. "Committee and Informal Leaders in the House of Representatives." Paper prepared for the Congressional Research Service-Dirksen Congressional Center Congressional Leadership Research Project and for the Annual Meeting of the Midwest Political Science Association, Chicago, April.

Hershey, Robert D. 16 July 1989. "Capitol Hill's High-Tech Tutor." *New York Times*: Section 3: 1, 12.

Kuntz, Phil. 27 August 1989. "Cautious Lawmakers Fret Over Biotech Issues." *Congressional Quarterly Weekly Report*: 2427–2430.

Lambrecht, Bill. 14 January 1990. "Expert Questions Safety of Hormone." *St. Louis Post Dispatch*: 7A.

Larrabee, John. 21 July 1989. "New Milk Production Drug Runs into Strong Opposition." *USA Today*, 3A.

Leahy, Patrick. Fall 1988. "Toward a National Biotechnology Policy." *Issues in Science and Technology*: 26–29.

McCormick, Douglas. August 1988. "Is Our Money Where Our Mouth Is?" *Bio/Technology* 6: 855.

Plein, L. Christopher. 1989. "The Emergence of the Pro-Biotechnology Coalition: Issue Development and the Agenda Setting Process." Paper delivered at the Annual Meeting of The American Political Science Association, Atlanta, August.

———. 1990. "Interest Group Coalition and Fracture: The Case of Biotechnology Opposition Groups." Paper Presented at the Annual Meeting of the Southern Political Science Association, Atlanta, Georgia.

———. forthcoming. "Popularizing Biotechnology: The Influence of Issue Definition." *Science, Technology, and Human Values*.

Plein, L. Christopher, and David J. Webber. Autumn 1988. "Congressional Consideration of Biotechnology." *Policy Studies Journal* 17: 75–86.

———. 1989. "Biotechnology and Agriculture: An Evolving Congressional Policy Arena." *The Political Economy of U.S. Agriculture: Choices for the 1990s*, edited by Carol Kramer. Washington, D.C.: Resources for the Future: 179–200.

Reppert, Barton. 5 January 1988. "OTA Emerges as Nonpartisan Player." *Washington Post*: A17.

Schneider, Keith. 12 January 1990. "F.D.A. Accused of Improper Ties in Review of Drug for Milk Cows." *New York Times*: 21.

Shapiro, Sidney. 5 September 1989. "Biotechnology and the Design of Regulation." First Draft of Report for the Administrative Conference of the United States.

Sinclair, Ward. 23 June 1986. "Biotechnology and the Milk Glut." *Washington Post*. p. 23.

Taylor, Sarah E. 1988. "Patenting Life." Issue Brief (Washington D.C., Congressional Research Service.

U.S. Congress. General Accounting Office. 1986. *Biotechnology: Analysis of Federally Funded Research*. Washington D.C.: USGPO.

———. 1988. *Biotechnology: Managing the Risks of Field Testing Genetically Engineered Organisms*. Washington D.C.: USGPO.

———. 1989. *Biotechnology: Backlog of Patent Applications*. Washington D.C.: USGPO.

U.S. Congress. House. 1980. *Genetic Engineering, Human Genetics, and Cell Biology: Evolution of Technological Issues, Biotechnology*. Prepared for the Subcommittee on Science, Research and Technology, Committee on Science and Technology by the Science Policy Research Division, Congressional Research Service. 96th Con. 2nd Sess., August.

———. 1985. *Planned Releases of Genetically-Altered Organisms: The Status of Government Research and Regulation*. Hearing before the Subcommittee on Investigations and Oversight of the Committee on Science and Technology. 99th Con. 1st Sess., December 4.

———. 1985. *Biotechnology and Agriculture*. Hearing Before the Subcommittee on Investigations and Oversight of the Committee on Science and Technology. 99th Con. 1st Sess., April 16–17.

———. 1986. *Issues in the Federal Regulation of Biotechnology: From Research to Release*. Report Prepared by the Subcommittee on Investigations and Oversight. Committee on Science and Technology. 99th Con. 2nd Sess. December. ———

———. 1986. *Review of Status and Potential Impact of Bovine Growth Hormone*. Hearing before the Subcommittee on Livestock, Dairy, and Poultry of the Committee on Agriculture. 99th Con. 2nd Sess., June 11.

———. 1986. *Ice Minus: A Case Study of the EPA's Review of Genetically Engineered Microbial Pesticides*. Hearing before the Subcommittee on Investigations and Oversight of the Committee on Science and Technology. 99th Con. 2nd Sess. March 4.

———. 1986. *USDA Licensing of a Genetically Altered Veterinary Vaccine*. Joint Hearing before the Subcommittee on Investigations and Oversight of the Committee on Science and Technology and the Subcommittee on Department Operations, Research, and Foreign Agriculture of the Committee on Agriculture. 99th Con. 2nd Sess., April 29.

———. 1986. *Coordinated Framework for the Regulation of Biotechnology*. Joint Hearing before the Investigations and Oversight Subcommittee; Natural Resources, Agricultural Research and Environment Subcommittee; and the Science, Research and Technology Subcommittee of the Committee on Science and Technology. 99th Con. 2nd Sess. July 23.

———. 1987. *Patents and the Constitution: Transgenic Animals*. Hearings before the Subcommittee on Courts, Civil Liberties, and the Administration of Justice of the Committee on the Judiciary. 100th Con. 1st Sess. June 11, July 22, August 21, and November 5.

———. 1988. *The Biotechnology Competitiveness Act*. Hearing Before the Subcommittee on Natural Resources, Agricultural Research and Environment and the Subcommittee on Science, Research and Technology of the Committee on Science, Space, and Technology. 100th Con. 2nd Sess., H.R. 4502 and S. 1966.

———. 1988. *Field Testing Genetically-Engineered Organisms*. Hearing before the Subcommittee on Natural Resources, Agricultural Research and Environment of the Committee on Science, Space, and Technology. 100th Con. 2nd Sess., May 5.

———. 1988. *Backlog of Patent Applications at the U.S. Patent and Trademark Office and Its Effect on Small High-Technology Firms*. Hearings before the Subcommittee on Regulation and the Business Opportunities of the Committee on Small Business. 100th Con. 2nd Sess., March 29.

U.S. Congress. Senate. 1976. *Oversight Hearing on Implementation of NIH Guidelines Governing Recombinant DNA Research*. Joint Hearing before the Subcommittee on Health of the Committee on Labor and Public Welfare and the Subcommittee on Administrative Practice and Procedure of the Committee on the Judiciary. 94th Con. 2nd Sess., September 22.

———. 1986. *Releasing Genetically Engineered Organisms into the Environment*. Hearing before the Subcommittee on Toxic Substances and Environmental oversight of the Committee on Environment and Public Works. 99th Con. 2nd Sess., May 16.

————. 1989. *The Human Genome Initiative and the Future of Biotechnology*. Hearing before the Subcommittee on Science, Technology, and Space of the Committee on Commerce, Science, Technology and Space of the Committee on Commerce, Science, and Transportation. 100th Con. 1st Sess., November 9.

U.S. Department of Agriculture. 1988. *Agricultural Biotechnology and the Public*. Proceedings Summary.

U.S. Department of Agriculture. Economic Research Service. 1987. *BST and the Dairy Industry: A National, Regional, and Far-level Analysis*.

Volger, David J. 1988. *The Politics of Congress*. Dubuque, Iowa: WM. C. Brown Publishers.

Webber, David J. 1990a. "Biotechnology Policy Knowledge: A Challenge to Congressional Policy-makers and Policy Analysts." In *Biotechnology: Assessing Social Impacts and Policy Implications*, edited by David J. Webber, 199–210. Westport, CT: Greenwood Press.

————. 1990. "The Emerging Federalism of Biotechnology Policy." Prepared for delivery at the 1990 Annual Meeting of the American Political Science Association, San Francisco Hilton, August 30–September 2, 1990.

Webber, David (ed.) 1988. "Symposium on Biotechnology, Agriculture, and Public Policy" *Policy Studies Journal* 17(1): 63–214.

————. (ed.) 1990. *Biotechnology: Assessing Social Impacts and Policy Implications*. Westport, CT: Greenwood Press.

Willard, Timothy and Daniel M. Fields. May–June 1989. "Helping Congress Look Ahead." *The Futurist*: 23–27.

Woodman, William E., Mack C. Shelley II, and Brian J. Reichel. 1989. *Biotechnology and the Research Enterprise: A Guide to the Literature*. Ames: Iowa State University.

9

Privacy, Efficiency, and Surveillance: Policy Choices in an Age of Computers and Communication Technologies

Priscilla M. Regan

Introduction

In the 1960s, computerization of personal information in the files of large organizations, such as credit agencies, banks, and the federal government, generated concern with the privacy rights of individuals. Many feared that with computerization more information would be collected and retained and then disclosed to and exchanged with other organizations. They believed that the storage capacity, speed of retrieval, and searching capabilities of computers would fundamentally change relationships between individuals and organizations. With computerization individuals would lose their privacy—their ability to control information about themselves.

Initially, members of Congress proposed comprehensive legislation to regulate the manual and computerized information practices of both public and private organizations. Quickly, however, the problem was divided up as different committees and subcommittees considered the privacy threats and the need for legislation for the organizations that were under their jurisdiction. For example, credit information came under the purview of one committee, information in government files another committee, and medical information yet another. Although the technological capabilities and problems were similar regardless of the type of personal information or the context in which it was being used, the congressional responses varied.

Congress did pass a number of laws protecting the privacy of personal information including the Fair Credit Reporting Act of 1970, the Family

Educational Rights and Privacy Act of 1974, the Privacy Act of 1974, the Tax Reform Act of 1976, the Right to Financial Privacy Act of 1978, and the Cable Communication Policy Act of 1984. In most cases, there was a critical event that focused attention on the issue and opened the "policy window" (Kingdon 1984) for serious policy formulation and adoption. For example, the 1974 Privacy Act was in part a response to the misuses of government information revealed during Watergate; the 1978 Right to Financial Privacy Act was a response to the Supreme Court's ruling in *Miller v. United States* (1976) that bank records were the property of the bank and that the individual had no property interest in those records; the current attempt to strengthen the 1970 Fair Credit Reporting Act is attributed to the *Business Week* report that one of its reporters easily gained access to the credit history of the Vice President; and the 1988 Video Privacy Act followed the *City Paper's* publication of the titles of the videos rented by Robert Bork, then a nominee for the Supreme Court.

Despite this privacy legislation, organizations have increased their use of computers and telecommunications for processing personal information. In the 1980s, the linking of separate computerized databases through online telecommunication lines provoked concern about the privacy rights of individuals—similar to the concerns of the 1960s. In order to detect instances of possible fraud or double-dipping, government agencies increasingly share personal information through techniques such as computer matching and front-end verification.[1] In the private sector, frequent shopper programs retain computerized databases on the buying habits of millions of consumers and then sell the information to marketing firms for targeted advertising. The merging of separate databases and the creation of what the congressional Office of Technology Assessment termed a *de facto* national database (U.S. OTA 1986, 3) call into question the effectiveness of the "first generation" of privacy laws. The continuing technological changes point to the problems with de-emphasizing the technological component and with legislating on a sector-by-sector, or recordkeeper-by-recordkeeper, basis.

Privacy and Technology

Langdon Winner distinguishes between science—"a particular way of knowing or body of knowledge"—and technology—"a particular kind of practice" (Winner 1977, 63). This is a case of technology and social change—how a new practice, computerization of record systems containing personal information, changes the relationship of the individual to organizations, especially how it changes how much the organization knows about and how much control it can exercise over the individual.

Emmanuel Mesthene's discussion of technology and social change is useful in analyzing the course of policy making under study here. Initially, computerization was just considered a *new way* of doing things that had been done before; later it became clear that computerization created *new possibilities* that did not exist before (Mesthene 1981, 56–59). Mesthene points out that technology creates new possibilities or opportunities which then require new organizations of human effort to realize and exploit them. This second-order effect involves social change. Policy choices can be made at either the first stage—what possibilities or opportunities should be pursued, or the second stage—responding to the social consequences of technological changes. Policy can then be proactive, shaping the course of technology change, or reactive, responding to technological changes. In the case of privacy and computers, congressional policy has been largely reactive.

Regardless of whether policies are proactive or reactive, policy decisions entail choices between values because as technology enables things to be done differently or makes it possible to do different things, it is likely, if not inevitable, that the configuration of interests and goals will change. As Mesthene points out:

> By making available new options, new technology can, and generally will, lead to a restructuring of the hierarchy of values, either by providing the means for previously unattainable ideals within the realm of choice and therefore of realizable values, or by altering the relative ease with which different values can be implemented—that is by changing the costs associated with them (Mesthene 1981, 63).

In this case, the new technology makes it easier, at least arguably, to achieve the value of organizational efficiency while making it more difficult to protect individual privacy. The policy choice then, at one level, is between the two values of privacy and efficiency. This is not a technical question, but a normative one. But, as Duncan MacRae observes, "the greatest problem which science poses for democracy, therefore, is that of the critical, logical examination of values which define democracy" (MacRae 1981, 508).

Mesthene and MacRae's emphasis on the value choices that occur in technology policy is mirrored by Harvey Brooks, who speaks of technology as "sociotechnical rather than technical." The social and managerial changes necessary to support technological innovations may "entail a sacrifice of other values that are too cherished in advanced industrial societies to be given up willingly" (Brooks 1981, 35, 52). The example Brooks cites is the infringement on civil liberties that would be necessary to ensure the safe and secure management of nuclear power. Similarly, the infringement on civil liberties—in this case, privacy—that would be necessary to have a totally efficient government administration may not be tolerable.

The policy issue for Congress is: when organizations use new computer and communication technologies to process personal information, what is the appropriate balance between privacy and efficiency? As we will see, three questions have recurred throughout the thirty years of policy discussion in Congress. As in other science and technology policies, these questions are a mixture of factual and ethical disagreements (Roberts, Thomas, and Dowling 1984, 112–122). The fact question is what changes are taking place as a result of organizational use of computers and telecommunications systems. The value question is how do these changes affect the balance between privacy and efficiency. The policy question is whether, and if so what, legislation is necessary. A brief examination of each of these questions may be instructive at this point.

The first recurring question in congressional policy making is to what extent did technology cause or contribute to invading privacy or upsetting the traditional balance between privacy and efficiency, and, therefore, to what extent should technology be the object of legislation. Over the thirty-year period of congressional policy making, innovations in computer and communication technologies have provided the catalyst for public concern and the backdrop for congressional legislation. The technological component that provoked the policy debate, and to a large extent drove support for legislation, was, however, not an essential component of the first generation of legislation. Initially, the problem was not defined in terms of computerization *per se*, but as one of how organizations use computerized records and how individual rights can be protected within that environment. Congressional understanding of the capabilities of computer and communication technologies and their expected future developments was not then the focus of initial congressional policy making. Instead, the focus was on understanding and getting concrete information on how organizations were using personal information. As technological innovations continued, policy makers realized that the organizational environment had qualitatively changed and that Congress needed to explicitly address the contribution that technology made to those changes.

The second question involves the definition of, and indicators for, the values of privacy and efficiency in order to evaluate whether there is a threat to privacy and/or a gain in efficiency. In the 1970s, the most commonly accepted definition of privacy was Alan Westin's—the right of individuals to control information about themselves (Westin 1967, 7). This was given somewhat more concrete meaning in the "Code of Fair Information Practices," which included the rights of individuals to see, correct, and amend personal information and included the principle that information collected for one purpose would not be used for another without the consent of the individual. As use of computer and communications technology in processing personal information accelerated, it became obvious that it was almost impossible for an individual to exercise *control* over information but an alternative definition of the privacy value was not offered. Instead privacy seemed to become more vague at the same time that efficiency became both politically and empirically clearer. Efficiency was more easily defined for policy purposes with the emphasis on the cost effectiveness of computer and communication systems in administering programs, especially in detecting instances of fraud, waste, and abuse.

The third policy question was what techniques of control were necessary and appropriate. Should individuals be given rights to protect their privacy? Should use of certain technologies be prohibited or certain organizations be given responsibilities and liabilities? Should the government regulate the personal information practices of public and private sector organizations? What congressional oversight was necessary to ensure that whatever policy it choose was implemented and effective?

In this chapter, we will examine how Congress responded to these questions in the case of federal government agencies' use of personal information. This case was chosen for a number of reasons. First, it provides the richest policy history for analysis; there are now two generations of laws, the Privacy Act of 1974 and the Computer Matching and Privacy Protection Act of 1988. Second, data about government information practices are, at least arguably, in the public domain; in the case of private sector information practices, proprietary concerns are often raised. Third, the case of government information practices raises interesting theoretical issues about the relationship of the individual to the government and the appropriate boundaries on government power.

This case of technology policy crosses a number of traditional policy arenas—individual rights, administrative practice and procedure, budgeting and appropriations, and technological change. Committees on the judiciary are unaccustomed to dealing with issues of science and technology and do not usually engage in cost-benefits analyses. Committees on governmental affairs and government operations also are not well versed in issues of science and technology. Science and technology committees are not generally confronted with issues of individual rights and administrative procedure. There are no incentives for members of budget and appropriations committees to take an active role in protecting values which are not easily quantifiable. As we will see, the main arenas for policy debate were the committees on governmental affairs and government operations.

1960s Literature of Alarm—Defining the Problem

For centuries public and private organizations have been collecting and maintaining information on individuals. For example, in the eleventh century, William the Conqueror compiled the Doomsday Book, filling it with information on each of his subjects in order to plan taxation and other state policies. Yet, the proper scope of information practices and its abuses was not a political issue until the 1960s. Two factors coalesced to bring the issue to the public agenda—an increase in record-keeping activities and the computerization of information processing.

Until the twentieth century, recordkeeping about individuals was limited to specific activities and was local in nature. As political, economic, and social relationships became more complex and organizations developed more formal arrangements for dealing with clients and customers, the nature of recordkeeping changed significantly. Organizations relied upon records to mediate their relationships with individuals. Accompanying this change was a qualitative increase in the sensitivity and personal nature of the information contained in these records. As the government expanded its social welfare activities, and as the economy became more credit oriented, and as society became more insurance conscious, more records containing information of a more sensitive and personal nature were maintained.

When records were held in local offices of a private company or government agency and consisted of paper in manila folders, the assumption was that the uses of the information were limited. With the computerization of large-scale recordkeeping in the late 1960s, concern about misuses of personal information took on a new intensity. In manual systems, if one wishes to retrieve a particular file or one specific item from a number of files, one has to physically retrieve each file and read through for the desired information—the "paperwork jungle" provides a physical limitation on the ability to search files. In computerized systems this takes a matter of seconds. The computer can also manipulate information in a way that time and memory prohibit someone from doing manually. For example, it can scan its files searching on the basis of certain characteristics or selecting for one individual. As one commentator notes, the computer makes possible "one-stop shopping" for all information about an individual (Project 1968). It can also sort, compare, and integrate various data files so that old data yield new information.

Data can also be exchanged speedily and cheaply through the telecommunications system. In pre-computer days, if one organization wanted information from another organization, the second organization had to transcribe the necessary information and mail or carry it to the first. If the first organization needed numerous files or was only interested in one item in the files, limitations of time and space could be prohibitive. With computer and telecommunication systems, such exchanges of data are routine and are done quickly and cheaply.

As the titles of the "literature of alarm" popular during this period indicate, the computer was characterized as signalling the dawn of "1984"— *The Privacy Invaders* (Brenton 1964), *The Intruders* (Long 1967), *The Naked Society* (Packard 1964), *Privacy and Freedom* (Westin 1967), *The Death of Privacy* (Rosenberg 1969), *The Assault on Privacy* (Miller 1971), and *On Record: Files and Dossiers in American Life* (Wheeler 1969). The computer was viewed as a threat to privacy because of its storage capacity, its speed in retrieving data, its ability to manipulate information, and its facility for transmitting information. These changes made it more difficult, as Alan Westin pointed out, for individuals to retain their privacy, "to determine for themselves when, how and to what extent information about them is communicated to others" (Westin 1967, 39).

By 1970, there was increasing public concern about the use of computers (Westin and Baker 1972, 465–485).[2] A survey conducted by the American Federation of Information Processing Societies (AFIPS) and *Time* magazine reported that at least 50 percent of the respondents believed that computerized files could be used to destroy individual freedom and that computers meant that "too many people have information about other people." Seventy-seven percent disagreed with the statement that "computers always give accurate information" and 42 percent agreed that "there is no way to find out if information about you that is stored in a computer is accurate." In addition, 84 percent of those interviewed thought that the government should be concerned about regulating the use of computers.

Congressional Response—Formulating Alternatives

Congressional interest in privacy and individual records was precipitated by the 1965 Social Science Research Council (SSRC) proposal that the Bureau of the Budget establish a Federal Data Center to provide and coordinate the use of government statistical information. In 1966, the Senate Committee on the Judiciary's Subcommittee on Administrative Practice and Procedure and the House Committee on Government Operations Special Subcommittee on Invasion of Privacy held hearings on this proposal. The chairman of the House subcommittee described the hearings as an attempt to find "a sense of balance" and to find out what information would be stored in the Federal Data Center, who would have access, and how confidentiality and privacy would be protected:

> It is our contention that if safeguards are not built into such a facility, it could lead to the creation of what I call "The Computerized Man." "The

Computerized Man," as I see him, would be stripped of his individuality and privacy. Through the standardization ushered in by technological advance, his status in society would be measured by the computer, and he would lose his personal identity. His life, his talent, and his earning capacity would be reduced to a tape with very few alternatives available (U.S. Congress, House 1966, 2).

As a part of this effort, the Senate subcommittee sent a questionnaire to all federal agencies to determine the amount and nature of personal information collected and how it was used by the agencies. This survey revealed that:

> In the mid-1960's, Federal files contained more than three billion records on individual citizens. Nearly one-half of these records were then retrievable by computer; they reportedly included over 27.2 billion names, 2.3 billion present and past addresses, 264.5 million criminal histories, 279.6 million mental health records, 916.4 million profiles on alcoholism and drug addiction, and over 1.2 billion financial records (U.S. Congress, Senate 1967a).

The study concluded that government agencies were asking citizens for too much information and that confidentiality provisions were non-existent or not meaningful (U.S. Congress, Senate 1967a, 8).

Following these congressional hearings, the Bureau of the Budget commissioned another study to explore ways of improving the storage of and access to government statistics. Again, the establishment of a National Data Center was recommended. In 1967 and 1968, both the House and Senate again held hearings and rejected the proposal for a National Data Center, in part because of the fear that once such a center was established it would be difficult to maintain its limited role and because Congress was not convinced that such a center would adequately protect the privacy of individual records.

In 1970, the Senate Judiciary Committee's Subcommittee on Constitutional Rights, chaired by Senator Ervin, began what became a four-year study of government databanks. These hearings were in response to the perceived fear of Americans that "the existing procedures are no longer sufficient to protect the privacy of the individual versus the 'information power of government'" (U.S. Congress, Senate 1971, 1). As part of this effort, the Subcommittee surveyed fifty-four federal agencies and found 858 databanks, noting that "there are without a doubt a great many more Federal data banks which the subcommittee, despite more than four years of patient effort, was unable to uncover" (U.S. Congress, Senate 1971, xxxvii). The subcommittee found that the majority of these databanks existed under unclear statutory authority and that government agencies failed to notify individuals that personal information about them was maintained in a databank.

At the same time that the Subcommittee on Constitutional Rights was conducting its study, two other advisory bodies were also investigating the effects of computerization on personal information systems. The Department of Health, Education and Welfare (HEW) set up a Secretary's Advisory Committee on Automated Personal Data Systems to analyze computerized information systems. In its 1973 report, *Records, Computers and the Rights of Citizens*, the HEW Committee identified three dangers resulting from the use of computerized recordkeeping: (1) an increase in organizational data processing capacity; (2) more access to personal data; and (3) the creation of technical recordkeepers (Department of Health, Education and Welfare 1973, 12). The HEW Committee emphasized the need for statutory protections because of the complex issues involving technology, individual rights and organizations. Despite rhetoric in support of a comprehensive policy, the final recommendations of the HEW Committee were rather weak, calling for the enactment of a Code of Fair Information Practice that would apply to both computerized and manual files.

In 1969, the Russell Sage Foundation and the National Academy of Science co-sponsored a project directed by Alan Westin and Michael Baker to examine organizational use of computers for personal information handling. Its 1972 report, *Databanks in a Free Society*, concluded that computerization of records was not the villain it was portrayed to be, and that the policy problem was not one of "information security to be resolved by technical specialists," but choices about "privacy, confidentiality and due process" (Westin and Baker 1972, 392). Again, the technological component of the issue was downplayed in favor of an emphasis on individual rights. Westin and Baker recommended a number of policy actions for both manual and computerized records including a "Citizen's Guide to Files," rules for confidentiality and data sharing, limitations on unnecessary data collection, technological safeguards, restricted use of the social security number, and the creation of information trust agencies to manage sensitive data.

By 1974, congressional committees, federal agencies, and private foundations had all contributed to the definition of the policy problem and to the formulation of alternatives. Their contributions did not vary in significant ways: there was agreement that the computer was the catalyst for the problem, but not the cause; that the problem was primarily one of protecting the privacy of individuals; that the problem existed in both the public and private sectors; and that some regulatory control had to be exercised over the ways in which organizations used personal information. The HEW Committee's Code of Fair Information Practice became both a summary of the way in which the problem was viewed and a blueprint for action. Its major principles included that:

> There must be no personal recordkeeping system whose very existence is secret.
> There must be a way for an individual to find out what information about him or her is in a record and how it is used.
> There must be a way for an individual to prevent information about him or her that was obtained for one purpose from being used or made available for other purposes without his or her consent.
> There must be a way for an individual to correct or amend a record of identifiable information about him or her.
> Any organization creating, maintaining, using, or disseminating records of identifiable personal data must assure the reliability of the data for their intended use and must take precautions to prevent misuse of the data (U.S. Department of Health, Education and Welfare 1973).

By this point in the policy process, the technological component—the computerization of records—had been defined out of the policy problem. Instead, the problem was defined as one of the bureaucratic information practices and individual privacy.

The Harder Step—Adopting a Meaningful Policy

In 1974, hearings on a number of privacy bills were held in both the Senate and the House. In the Senate, the Committee on Government Operations and the Judiciary Committee held joint hearings on five privacy bills (U.S. Congress, Senate 1967b). The major bill[3] on which debate focused was S. 3418, introduced by Senators Ervin, Percy and Muskie, which covered all automated and manual personal information systems in federal, state, and local governments, as well as the private sector. This bill provided for a federal Privacy Board with regulatory authority to enter premises where information was held and by subpoena compel the production of documents, to hold hearings regarding violations and exemptions, and to order an organization to cease and desist unauthorized information practices. It also established fair information practices, similar to the HEW Code, and gave individuals the rights to see and amend their files and to be informed of dissemination of information. The House Committee on Government Operations also held hearings in 1974 on a Privacy Bill, H.R. 16373, which was less inclusive and provided weaker protections than the Senate bill: it covered only federal agencies, did not provide for an independent privacy board, and allowed for more exemptions (U.S. Congress, House 1974).

The two most contentious issues in congressional hearings were whether the public and private sectors should be subject to similar legislation, and whether a Privacy Board should be established. In both cases, those opposed—federal agencies and private sector organizations—were able to weaken the scope of the proposed legislation and to defer these policy decisions. To a large extent, the adoption of the Privacy Act constituted more symbolic than real support for privacy. Many members of Congress believed that some formal recognition of privacy interests was necessary following the revelations of government abuses of personal information during Watergate. The 1974 Privacy Act that finally passed both houses reflected the original House bill in that it covered only federal agencies and did not provide for a separate agency to oversee agency information practices.

The goal of the Privacy Act was "to provide certain safeguards for an individual against an invasion of personal privacy." In justifying the need for legislation, Congress in section 2 of the Privacy Act, concluded that:

> The increasing use of computers and sophisticated information technology, while essential to the efficient operations of the Government, has greatly magnified the harm to individual privacy that can occur from any collection, maintenance, use or dissemination of personal information (U.S. Congress, House 1974).

This is the only mention of technology in the act. Instead, the object of legislation is "records" and "systems of records," both of which are defined in terms of paper files. To protect privacy, the Privacy Act gives individual rights of access, correction, and knowledge about personal records in computerized or manual files, and subjects federal agencies to standards of fair information handling, including that personal information must be accurate, complete, relevant and timely, and used only for the purpose for which it is collected.

The congressional scheme for implementing and enforcing these rights and responsibilities is decidedly weak: OMB was given responsibility for implementation and oversight of the Act; individuals were given the right to bring civil suits to enforce agency compliance; and the Privacy Protection Study Commission was established to investigate the need for legislation over the private sector and for an oversight body over federal agencies. In 1977, the Commission reported that the Privacy Act had not resulted in the public benefits expected, in part because of its ambiguity and in part because of its failure to address the technological changes. It recommended the establishment of a Federal Privacy Board with responsibilities similar to those approved by the Senate in 1974 (PPSC, *Personal*, 1977, 37).

The Late 1970s—Oversight and Contradictions

Following passage of the Privacy Act and the recommendations of the Privacy Protection Study Commission, congressional interest in the issue waned. The judiciary and governmental affairs committees conducted some superficial oversight and occasionally a bill was introduced to strengthen the Privacy Act, usually by establishing a permanent privacy board. Generally, however, congressional interest in privacy was replaced by congressional and executive interest in the efficiency and effectiveness of governmental programs. The terms of the debate shifted from concern about computers and privacy to an emphasis on computers as a tool in detecting fraud, waste, and abuse. The policy arenas in which these issues were debated shifted within Congress from judiciary and individual rights committees and subcommittees to management and budget committees and subcommittees and on the whole shifted from Congress to the executive, especially to various presidential commissions such as the President's Private Sector Survey on Cost Control (the Grace Commission) and the President's Council on Integrity and Efficiency.

The Privacy Act was compromised and weakened by a series of executive and legislative actions. The first of these was Project Match, initiated by HEW Secretary Califano in November 1977, which compared the computerized files of federal employees with the computerized files of those receiving benefits through Aid to Families with Dependent Children (AFDC) in order to detect those government employees who were fraudulently receiving AFDC benefits (Kirchner 1981, 1–16). Privacy advocates in Congress, members of the Privacy Protection Study Commission, the ACLU, and others criticized the proposed match as a "fishing expedition." Some federal agencies, including the Civil Service Commission and the Department of Defense questioned whether the Privacy Act prohibited such a use of their files because the use was not compatible with the purpose for which it was collected and hence required individual consent. Despite these criticisms and hesitations, the match was done and by March 1978 Project Match had identified 7,100 employees who were *possibly* ineligible for AFDC benefits and had generated so much information that agency officials could not follow up to determine that validity of that information (Weiss 1983, 432).

In response to congressional, presidential, and interest group concerns about the privacy implications of Project Match, OMB—with assistance from the President's Office of Telecommunications Policy and the White House Privacy Initiative—was given responsibility for writing guidelines for computer matching. At this point policy deliberations retreated deep inside the executive branch. The OMB guidelines, issued in March 1979, allowed computer matches to occur as a "routine use" exemption to the

Privacy Act[4] and required any disclosures of personal information during a match to be made in accordance with the routine use limitations. (i.e., notices in the *Federal Register* prior to the match with time for public comment). Under the OMB guidelines (Office 1979), a match was to be performed only if there was a "demonstrable financial benefit," and, to this end, the guidelines required documentation of benefits, costs, potential harm, and alternatives considered to detect or curtail fraud and abuse. The guidelines also required agencies to submit a report describing the match to the Director of OMB, the Speaker of the House, and the President of the Senate. The purpose of the guidelines was "to aid agencies in balancing the government's need to maintain the integrity of Federal programs with the individual's right to personal privacy" (OMB 1979). In theory, the guidelines were capable of achieving that goal, but in practice, agencies did not follow the guidelines, OMB did not monitor agencies activities, the public and its interest groups did not comment on *Federal Register* notices, and there was little congressional reaction. In this policy vacuum, agency use of computer matching increased.

During this same time period, Congress passed a number of laws whose purpose was either to reduce fraud, waste and abuse in government programs or to increase the overall efficiency of governmental programs. Each of these contributed to a political environment that condoned the sharing of personal information. The laws establishing inspectors general offices in a number of federal agencies—including Health and Human Services, Labor, and Defense—emphasized the public policy goal of detecting fraud, waste, abuse, and management deficiencies and gave to these offices discretionary power in initiating audits and investigations and in choosing techniques to meet these goals. Computerized matching and other record searches were primary techniques in these efforts. A second legislative endeavor that is perceived as encouraging data sharing among federal agencies is the Paperwork Reduction Act of 1980 which gave OMB oversight authority for federal agency information practices and the responsibility to promote the effective use of information technology. The Federal Managers Financial Integrity Act of 1982, which required periodic evaluations of and reports on agency systems of internal control and action to reduce fraud, waste, abuse, and error, may also encourage information sharing.

Two statutes, the Debt Collection Act of 1982[5] and the Deficit Reduction Act of 1984,[6] specifically require exchanges of personal information and encourage the use of computers and telecommunications in these exchanges. Although both laws provide certain due process rights for the individual, the use of personal information for a purpose other than that for which it was collected is legitimated. In addition to these broad endorsements of and requirements for computer matches, there are a number of statutes that authorize specific computer matches, including the Tax Reform Act of 1976, the Food Stamp Act Amendments of 1980, and the Department of Defense Authorization Act of 1983.[7]

While these various information exchanges were being legitimated, three congressional committees—Senate Committee on the Judiciary, House Committee on the Judiciary, and the House Post Office and Civil Service Committee—asked the congressional Office of Technology Assessment (OTA) to assess the societal impacts of national information systems. This request reflected renewed congressional interest in the effects of computerization of information systems. The first report issued in 1981, *Computer-Based National Information Systems*, provided a survey of recent developments in computer and communication technologies and a framework for understanding the policy issues that result. Included among these issues was privacy, with questions raised as to the effectiveness of the existing sector-by-sector approach (U.S. OTA 1981, 73–78).

Beginning in 1982, the congressional committees with jurisdiction over the Privacy Act re-entered the discussion. In 1982, the Subcommittee on Oversight of Government Management of the Committee on Governmental Affairs held hearings on computer matching (U.S. Congress, Senate 1983, 1). Senator Cohen, chairman of the Subcommittee, in his opening remarks, emphasized the congressional oversight responsibility "to determine whether the Government is striking a proper balance between the important concerns of Government efficiency and individual privacy" (U.S. Congress, Senate 1983, 1). Two panels of witnesses testified at this hearing—one of privacy experts, and one of inspectors general and members of the President's Council on Integrity and Efficiency. In 1983, the Government Information, Justice, and Agriculture Subcommittee of the Committee on Government Operations held hearings (U.S. Congress, House 1983). Representative English, noted that these were the first such hearings on the Privacy Act, and that they were precipitated, in part, by his concern "that the bureaucracy has succeeded in avoiding most of the act's substantive limitations on the use of information" (U.S. Congress, House 1983, 2).[8]

Both congressional hearings were important in establishing that there was a problem with the Privacy Act itself and OMB enforcement thereof. However, the dimensions of the problem and the alternatives for dealing with that problem still needed to be determined. To this end, the General Accounting Office, initially at the request of some members of Congress and later on its own initiative, began a number of studies of the use of computer matching by federal and state agencies and also of the implementation of the Privacy Act. From 1976 to 1984, GAO conducted ninety-two reports on eligibility verification programs at the state and federal levels, and fifty-six reports relating to privacy and information systems

(U.S. Congress. GAO 1985). Additionally, in 1983, the Senate Committee on Governmental Affairs and the House Committee on the Judiciary's Subcommittee on Courts, Civil Liberties and the Administration of Justice requested that OTA examine federal agency use of new information technologies and evaluate their implications both for congressional oversight of agency activities and for individual rights.

The Reality of the 1980s—Reassessing the Technology and the Problem

The increase in the use of computer matching reawakened privacy advocates inside and outside the government, and reopened the debate on the definition of this policy problem. At this time, the policy concern shifted from one of "privacy" to one of "surveillance." The capabilities that computerized databases offered and the merging of separate databases made it possible for the organizations using and maintaining the databases to monitor the present and past activities of individuals. This aspect of computerized information systems had been recognized in the early debates. For example, in 1973 sociologist James Rule in *Private Lives and Public Surveillance* argued that:

> The commonest solutions to the problem of mass surveillance in large-scale societies lie in the use of documentation . . . the formal rendering of information about people comes to take the place of informal mechanisms of surveillance found in small-scale settings. The crucial function of such documentation is to link people to their pasts, and thereby to provide the surveillance necessary for the exercise of social control (Rule 28).

As the use of computer matching and other record searches increased, the scholars and journalists within this policy community began to emphasize the surveillance aspect of the problems caused by computers. David Burnham's *Rise of the Computer State* adopted this perspective, arguing that, for instance, the computerized databanks of credit agencies or car rental agencies contained records on each transaction of each customer, and that "without such general surveillance operations by American business, of course, the quick access to credit and the mobile life our society prizes would be impossible" (Burnham 1983, 39). Similarly, Gary Marx and Nancy Reichman suggested that computers were "Routinizing the Discovery of Secrets," in that systematic data searching "permits the joining of heretofore independent pieces of information in order to expose offenses and offenders that would remain hidden unless such links could be drawn" (Marx and Reichman 1984, 423).

With the arrival of the year "1984," the surveillance theme received more attention both within Congress and outside. The Subcommittee on Courts, Civil Liberties, and the Administration of Justice of the House Judiciary Committee held a series of hearings on "1984" and the National Security State (U.S. Congress, House 1984). Senator Cohen's Subcommittee on Oversight of Government Management held hearings on the computer matching of taxpayer records (U.S. Congress, Senate 1984). With this emphasis on surveillance, the technological capabilities of computer and communications technologies again became important issues in the policy debates.

A number of interest groups also organized conferences around this theme. The American Bar Association (ABA) in August of 1984 held a symposium on "Information Law and Ethics: In the Shadow of Orwell— The Citizen and Government." The American Civil Liberties Union (ACLU) and the Public Interest Computer Association (PICA) held two consultations, in June of 1984 and January of 1985, to explore privacy issues posed by new computer and communications technologies. Congressional staff from both the House and Senate were participants in both the ABA and the ACLU/PICA conferences.

Consistent again with the theme of "1984," Louis Harris & Associates conducted a poll to ascertain public attitudes toward the new technologies and their impact on American life (Louis Harris 1984).[9] In response to the question "How concerned are you about threats to your personal privacy in America today?", 48 percent of the respondents described themselves as "very concerned" and 29 percent as "somewhat concerned." This reflected a significant increase in concern with privacy from polls done in 1978 in which 25 percent described themselves as "very concerned." The percentage of the public believing computers to be a threat to privacy increased from 38 percent in 1974 to 51 percent in 1983. The survey also revealed that an increasing percentage of the public did not believe that the privacy of personal information in computers was adequately safeguarded—from 52 percent in 1978 to 60 percent in 1983—and that 68 percent of the 1983 respondents believed that the use of computers must be sharply restricted in the future if privacy is to be preserved. Most respondents, 84 percent, believed that master files containing personal information—credit and employment histories, medical history, phone calls, buying habits, and travel—could be compiled "fairly easily." Seventy-eight percent believed that if such a master file were put together, it would violate their privacy.

In the 1983 Harris survey, strong majorities supported the enactment of new federal laws to deal with information abuse including laws that would require that any information from a computer that might be damaging to people or organizations must be double-checked thoroughly before being used and laws that would regulate what kind of information about an individual could be combined with other information about the same individual. Although 62 percent thought it was very important that there be an independent agency to handle complaints about violations of personal privacy by organizations, 46 percent were opposed to the creation of a National Privacy Protection Agency. The authors of the Harris analysis observed that:

> Particularly striking is the pervasiveness of support for tough new ground rules governing computers and other information technology. Americans are not willing to endure abuse or misuse of information, and they overwhelmingly support action to do something about it. This support permeates all subgroups in society and represents a mandate for initiatives in public policy (Louis Harris 1983, 41).

Following release of this public opinion survey, the House Subcommittee on Government Information, Justice and Agriculture held hearings in which a number of witnesses testified that the Harris survey documented both a need for stronger privacy legislation and an increase in the public's support for such legislation (U.S. Congress, House 1984).[10]

In response to the need for information on how agencies were using new technologies and with what effects, OTA conducted a survey of federal agencies to gather data on a number of topics including Privacy Act implementation, computer matching and front-end verification, third-party information and profiling, use of credit reports, and electronic records management and electronic mail. The survey was sent to thirteen cabinet-level agencies and twenty selected subcabinet agencies; a total of 142 agency components provided information. Among the findings of this survey were the following:

> respondents reported 539 Privacy Act record systems with 3.5 billion records, 42 percent of these record systems were fully computerized;
> although agencies responding reported an increase from a few thousand microcomputers in 1980 to about 100,000 in 1985, only about 8 percent had revised or updated their Privacy Act guidelines with respect to microcomputers;
> only about 12 percent of agencies reported that they had conducted record qualities audits;
> from 1980 to 1984, the number of computer matches done by federal agencies nearly tripled;
> the number of separate records used in the reported matching programs totaled over 2 billion; the total number of records matched was reported to be over 7 billion due to multiple matches of the same records;

43 percent of agency components that reported participation in computer matching activities (16 out of 37) said that the matches were required or authorized by legislation; and

68 percent (25 of 37) said that procedures were used to ensure that the subject record files contained accurate information (U.S. OTA June 1986).

These data were important in documenting the scope of the policy problem and the changes that had taken place in federal agency practices as a result of innovations in computers and telecommunications. OTA's analysis of federal agency use of electronic record systems—specifically for computer matching, front-end verification, and computer profiling—revealed four common policy problems:

1) New applications of personal information had undermined the goal of the Privacy Act that individuals be able to control information about themselves.

2) There was serious question as to the efficacy of the current institutional arrangements for oversight of federal agency compliance with the Privacy Act and related OMB guidelines.

3) Neither Congress nor the executive branch was providing a forum in which the privacy, management efficiency, and law enforcement implications of federal electronic record system applications can be fully debated and resolved.

4) Within the federal government, the broader social, economic, and political context of information policy, which included privacy-related issues, was not being considered (U.S. OTA June 1986, 4–6).

OTA concluded that the widespread use of computerized databases, electronic record searches and matches, and computer networking was leading rapidly to the creation of a *de facto* national database containing personal information on most Americans and that the use of the social security number as a *de facto* electronic national identifier facilitated the development of this database (U.S. OTA June 1986, 3).

Congressional Response—Incremental Change or Rethinking?

Armed with the new data and analysis conducted by OTA, the GAO studies, and the Harris public opinion surveys, the Senate Committee on Governmental Affairs took up the issue. In 1986, Senators Cohen and Levin introduced the "Computer Matchings and Privacy Protection Act" (S.2756) to ensure privacy, integrity, and verification of data disclosed for computer matching, and to establish Data Integrity Boards within federal agencies. Testifying at these hearings were representatives from OMB, GAO, OTA, ACLU, and the ABA. The two controversial parts of the bill were the matching agreement required before a match was to take place and the establishment and powers of the Data Integrity Board. All witnesses, except OMB, gave fairly strong support to the Data Integrity Boards. OMB noted that the bill "clearly authorizes the government's use of computer matching," (47) but questioned whether "adding another bureaucratic layer to agency level oversight is appropriate" (U.S. Congress, Senate 1986, 71).[11] The language and position of OMB were almost identical to that of federal agencies in opposition to the original proposals for a Privacy Act.

Following these hearings, the Subcommittee revised and approved a revised Computer Matching and Privacy Protection Act (S.496), which was approved by the full Senate in May 1987. A subcommittee of the House Committee on Government Operations held one day of hearings at which GAO, ACLU, ABA, and OMB again appeared as witnesses (U.S. Congress, House 1981). In October 1988, a revised version of the Computer Matching and Privacy Protection Act was passed by both houses and signed by the President.

To some extent the Computer Matching and Privacy Protection Act represents an incremental change. Computer matching is legitimated by statute, although in keeping with the Privacy Act procedural requirements are placed on the agencies and individual rights are given some protection. For the first time, however, legislation makes a distinction based on technology and directly addresses the technology. Also for the first time, legislation establishes some agency-wide, although not government-wide, oversight over agency matching activities. Until this time, the inspectors general, operating largely through the President's Council on Integrity and Efficiency and OMB, controlled agency matches. OMB resisted the establishment of the Data Integrity Boards and was able to delay their establishment because, as Representative Wise noted in introducing a bill to extend the date by which agencies were to establish Data Integrity boards, "OMB took longer than it should have to issue implementing regulations required by the Act."[12] As with the Privacy Act itself, the subsequent powers of the Data Integrity Board are likely to depend on congressional oversight, either through the relevant subcommittees or through outside investigations.

Conclusion

Although the innovations in computer and communication technologies provided the catalyst for public concerns and the backdrop for congressional policy formulation, the technology did not drive the policy decisions. Instead, Congress initially dealt with this as a case of administrative procedures, where policy establishes rules for mediating relationships between people and organizations. The problem was not defined in terms of computerization *per se*, but as one of how organizations use personal information and how individual rights can be protected within that environment. The emphasis on privacy and de-emphasis of technology contributed to ineffective legislation. But the congressional emphasis was consistent with the early policy analysis. The HEW Committee and the National Academy of Sciences and Russell Sage Foundation studies established the parameters of the policy debate. It was not until the 1980s, with widespread computer matching and the 1986 OTA study, that the scale and scope of technological changes were recognized.

The emphasis on protecting individual privacy deflected attention away from the real changes that were taking place in bureaucratic power as a result of new computer and communication technologies. Federal agencies and other organizations were then relatively free to pursue the benefits that they saw in these technologies. The changes in bureaucratic power took place without much public notice. James Beniger persuasively argues that the development and use of computer and communication technologies must be seen as part of a larger social and historical change that he calls the "Control Revolution" (Beniger 1986) in which bureaucracies and data processing play a central role. Given this perspective, giving individual rights to protect their privacy appears to be an inadequate response to changes of such magnitude. It is not that the goal of protecting privacy is inappropriate, but that the means to protect it are not sufficient.

The lack of definition or conceptual clarity for the value "privacy" also complicated policy efforts. The first sentence of Alan Westin's seminal work *Privacy and Freedom* is: "Few values so fundamental to society as privacy have been left so undefined in social theory or have been the subject of such vague and confused writing by social scientists" (Westin 1967, 52). Similarly, C. Herman Pritchett notes in the Preface to *Privacy, Law, and Public Policy* that "Privacy is a confusing and complicated idea" (O'Brien 1978, vii). These difficulties in conceptualizing privacy have carried over into difficulties in formulating policy to protect privacy.

Difficulties in conceptualizing fundamental values is not unique to this policy area. In discussing science and technology policy, Harrison observes that: "In the absence of well-defined and generally accepted systems of quality-of-life indicators, we have become dependent on economic indicators" (Harrison 1984, 123–125). Similarly, Tribe et al. distinguish between "hard" values and "soft" values and find that the analytical techniques used in environmental policy analysis were biased, "Thus 'hard' values, such as short-term economic efficiency, would be likely to swamp 'soft' values, such as ecological balance, and even 'softer' ones, such as the love of natural beauty" (Tribe 1976, x). In this case the "hard" values of government efficiency and the detection of fraud, waste and abuse prevailed over the "soft" value of privacy.

To improve policy making in this area, Congress needs to acknowledge the power that computers and telecommunications bring to an organization. With the merging of computer and communication technologies and the reality of a *de facto* national database, the congressional policy concern has shifted to the surveillance capabilities of *systems*, rather than the privacy threats from *records*. This perspective reflects a more realistic understanding of the policy problem and may signal a different congressional approach, one that may allow for more effective legislation.

One possible avenue through which fundamental rethinking on this issue might come is the interest in this issue recently expressed by the House Subcommittee on Telecommunications and Finance. This subcommittee, through its jurisdiction over the public and private telecommunications networks, has raised questions about the use of these networks to match personal information. It has asked GAO to investigate this area. It is possible that this approach could result in proposed legislation that targeted not agency or private sector uses of information, but their access to the telecommunications system for certain purposes (i.e., exchanges of personal information). If this was the case, it is likely that such legislation might be an amendment to the Electronic Communications Privacy Act and that such legislation would not distinguish between the public sector and the private sector.

Recent technological changes in organizational information practices give evidence for the need for such an approach. Organizations are increasingly using computer and communication technologies throughout all stages of information practices making it difficult to speak of separate records, separate record systems, or even separate organizations. For example, in eligibility benefit programs, a number of agencies are now using computer and communication technologies to distribute cash or in-kind benefits to recipients and to credit the provider for the value of a good or service. Once again, the efficiency proponents within the executive have supported these systems although the cost-effectiveness of these pilots has not been demonstrated (U.S. OTA 1988).

With an understanding of the policy problem that recognizes how computer and communications technologies are changing organizational practices, the weaknesses in an enforcement approach that relies on individual suits become obvious. A regulatory approach that allows for ongoing oversight and flexible responses to technological change then appear necessary. The establishment of an advisory or regulatory agency to oversee agency practices, technological developments, and individual rights would provide the continuous monitoring and flexibility that seems necessary. Such an agency was proposed in the original Privacy Act passed by the Senate in 1974, was recommended by the Privacy Protection Study Commission in 1977, and was considered as a viable policy option by OTA in 1986. Moreover, every other Western democracy has established either a Privacy Commissioner—Canada and Australia—or a Data Protection Authority—Britain, Germany, and Sweden—to provide oversight and flexibility in response to technological change (Flaherty 1989).

It may be useful, in conclusion, to step back and place this case in a broader context of congressional policy making. I have argued that congressional efforts have been ineffective in dealing with the problems presented by federal agencies' use of new computer and communications technologies. At the same time, however, it is somewhat surprising that Congress passed any legislation. This is an issue for which there is diffuse public support—in theory everyone wants his or her privacy protected, but in reality no one cares until some harm has occurred. Additionally, federal agencies, with the support of OMB and whatever administration was in power, have consistently been opposed to legislation and have actively lobbied against it. Their opposition has also been in keeping with the public concern in the 1980s on efficiency and deficit reduction. Within this context, it is surprising that Congress was even able to pass the largely symbolic Privacy Act of 1974, keep the issue on the congressional oversight agenda during the 1980s, and pass a stronger privacy measure in 1988. Credit here must be given to the policy community concerned with privacy and technology—the American Civil Liberties Union, the Individual Rights Subcommittee of the American Bar Association, and the Computer Professionals for Social Responsibility—and the staffs of the relevant congressional committees. Through their efforts, OTA and GAO studies were requested, congressional hearings were held, and conferences were convened. This policy activity kept the issue on the congressional agenda, tracked the new uses of computer and communication technologies, and posed an alternative to the executive branch's emphasis on efficiency.

Notes

1. Computer matching involves the comparison of two or more sets or systems of computerized records to search for individuals who may be included in more than one file. Front-end verification is used to check the accuracy and completeness of information that an individual supplies against information already held in a computerized database.

2. In 1970–1971, there were three separate national opinion polls that touched on this area. The 1970 Louis Harris poll included questions on privacy and privacy-related topics. A 1971 study of public attitudes towards computers was sponsored jointly by the American Federation of Information Processing Societies (AFIPS) and *Time* magazine. In 1970, the Retail Credit Company commissioned the Opinion

Research Corporation to conduct a national survey entitled "Investigations Relating to Insurance, Job and Credit Applications: A Nationwide Survey of Public Attitudes." There were also regional and special opinion surveys on these questions including: congressional polling on constituent opinions; a 1971 survey of college students conducted for the Rockefeller Foundation, the 1971 White House Conference on Youth, and a 1970 study on the public's perception of technology commissioned by the Harvard Program on Technology and Society.

3. The other four bills were: S.2633, introduced by Senators Ervin, Bayh, Goldwater, Kennedy, and Mathias, which covered only Federal personal information systems and those state and local systems funded by the Federal Government and gave oversight and registration authority to the General Accounting Office; S.2542, introduced by Senator Bayh, which covered only Federal Government records and provided for a Federal Privacy Board; S.2810, introduced by Senator Goldwater, which covered all public information systems and private systems that were supported by Federal funds; and S.3116, introduced by Senator Hatfield, which prohibited the sale or distribution of certain personal information.

4. Section 3b of the Privacy Act establishes the conditions under which an agency can disclose personal information to another party without the prior consent of the individual. One of these conditions of disclosure is for a "routine use," defined as "the use of such record for a purpose which is compatible with the purpose for which it was collected." [3(a) (7)] All routine uses are to be published in the *Federal Register*, including "the categories of users and the purpose of such use." [3(e) (4) (D)].

5. The Debt Collection Act of 1982 sets up a system of data sharing between federal agencies and private credit reporting agencies in order to increase the collection of delinquent nontax debts. Specifically, the act permits federal agencies to refer delinquent nontax debts to credit bureaus to affect credit ratings, require applicants for federal loans to supply their taxpayer identification numbers, screen credit applicants against IRS files to check for tax delinquency, and turn over to private contractors the mailing addresses of delinquent debtors obtained from the IRS. Rather than establishing a new government bureaucracy or giving additional powers to an existing one, OMB decided that "the existing nationwide network of commercial and consumer credit bureaus will be under contract to provide this service for all departments and agencies."

6. The 1984 Deficit Reduction Act requires the most far-reaching data sharing at both the state and federal levels. DEFRA requires states to establish information systems containing income and wage information from employers for purposes of verifying eligibility for benefit programs and to exchange relevant information with other state agencies and with HHS.

7. The Tax Reform Act of 1976 permits HEW to search the databases of other federal agencies in order to locate parents who were not keeping up with their child support payments. The Food Stamp Act Amendments of 1980 allow state food stamp agencies to use wage, benefit and other information in the files of the Social Security Administration and of state unemployment compensation agencies. The Department of Defense Authorization Act of 1983 requires the Secretary of Education to establish a method to verify that any individual receiving a federal grant or loan had complied with selective service registration requirements.

8. There had actually been an earlier oversight hearing, chaired by Representative Bella S. Abzug, soon after the Privacy Act went into effect. US Congress, Committee on Government Operations, Government Information and Individual Rights Subcommittee. *Implementation of the Privacy Act of 1974: Data Banks.* 94th Congress, 1st session. 3, June 1975. Washington, D.C.: GPO, 1975.

9. In 1979 a similar survey had been conducted by Louis Harris and Associates Inc. and Alan F. Westin, *The Dimensions of Privacy: A National Opinion Research Survey of Attitudes Toward Privacy.* Conducted for Sentry Insurance, Dec. 1979.

10. The same subcommittee had held hearings in 1979 when an earlier Harris survey, "The Dimensions of Privacy," was released, see: *Public Reaction to Privacy Issues,* Hearings, 96th Congress, 1st session. 6 June 1979. Washington, D.C.: GPO, 1980.

11. Statement of Joseph R. Wright, Jr., Deputy Director of OMB.

12. Mr. Wise remarks on the floor, July 11, 1989, *Congressional Record - House,* H3562.

Bibliography

Beniger, James R. 1986. *The Control Revolution.* Cambridge: Harvard University Press.

Brooks, Harvey. 1981. "Technology, Evolution, and Purpose." In *Science, Technology, and National Policy,* edited by Thomas J. Kuehn and Alan L. Porter. Ithaca: Cornell University Press.

Burnham, David. 1983. *The Rise of the Computer State.* New York: Random House.

Flaherty, David H. 1989. *Protecting Privacy in Surveillance Societies.* Chapel Hill: University of North Carolina Press.

Harrison, Anna J. 1984. "Commentary." *Science, Technology, and Human Values* 9(1): 123–125.

Kingdon, John W. 1984. *Agendas, Alternatives, and Public Policies.* Boston: Little, Brown and Company.

Kirchner, Jake. 14 December 1981. "Privacy—A History of Computer Matching in the Federal Government." *Computerworld.*

Louis Harris & Associates Inc. December 1983. *The Road After 1984: A Nationwide Survey of the Public and Its Leaders of the New Technology and Its Consequences for American Life.* Conducted for Southern New England Telephone for presentation at The Eighth International Smithsonian Symposium.

MacRae, Duncan Jr. 1981. "Science and the Formation of Policy in a Democracy." In *Science, Technology, and National Policy,* edited by Thomas J. Kuehn and Alan L. Porter. Ithaca: Cornell University Press.

Marx, Gary T., and Nancy Reichman. March/April 1984. "Routinizing the Discovery of Secrets: Computers as Informants." *American Behavioral Scientist,* 27(4): 423–452.

Mesthene, Emmanuel G. 1981. "How Technology Will Shape the Future." In *Science, Technology, and National Policy,* edited by Thomas J. Kuehn and Alan L. Porter. Ithaca: Cornell University Press.

O'Brien, David M. 1979. *Privacy, Law, and Public Policy.* New York: Praeger.

Office of Management and Budget. 30 March 1979. "Guidance to Agencies on Conducting Automated Matching Programs." Memo from James T. McIntyre, Jr., to the Heads of Executive Departments.

———. 15 December 1982. "OMB Announces Progress in Administration's Debt Collection Effort." OMB 82–32. (Reform '88 Communications). Government to Use Credit Bureaus to Cut Delinquent Debts; Delinquency Growth Halted, OMB 83–89 (Public Affairs/Management) Sept. 23, 1983.

Privacy Protection Study Commission. 1977. *Personal Privacy in an Information Society.* Washington, D.C.: GPO.

———. 1977. *The Privacy Act of 1974: An Assessment.* Appendix 4 to the Report of the Privacy Protection Study Commission. Washington D.C.: GPO.

Project. 1968. "The Computerization of Government Files: What Impact on the Individual?" *UCLA Law Review* 15: 1374, 1441.

Roberts, Marc J., Stephen R. Thomas, and Michael J. Dowling. Winter 1984. "Mapping Scientific Disputes that Affect Public Policymaking." *Science, Technology, and Human Values* 9(1): 112–122.

Rule, James B. 1973. *Private Lives and Public Surveillance.* London: Allen Lane.

Tribe, Laurence H., Corinee S. Schelling, and John Voss. 1976. *When Values Conflict.* Cambridge: Ballinger Publishing Company.

U.S. Congress. General Accounting Office. 1 March 1985. *Eligibility Verification and Privacy in Federal Benefit Programs: A Delicate Balance.* Washington D.C.: GPO.

U.S. Congress. House. Committee on Government Operations. Special Subcommittee on Invasion of Privacy. 1966. *The Computer and Invasion of Privacy.* Hearings, 89th Cong. 2nd sess., 26–28 July 1966.

———. Committee on Government Operations. 1974. *Privacy Act of 1974.* 93rd Cong. 2nd sess. pp. 93–1416.

———. Subcommittee on Government Information, Justice, and Agriculture Subcommittee. 1983. *Oversight of the Privacy Act of 1974.* 98th Cong. 1st sess.

———. 1984. *Privacy and 1984: Public Opinions on Privacy Issues.* Hearings, 98th Cong. 1st sess., 4 April 1984.

———. Committee on Judiciary. Subcommittee on Courts, Civil Liberties and the Administration of Justice. 1984. *1984: Civil Liberties and the National Security State.* Hearings. 98th Cong. 1st and 2nd sess. Nov. 2 and 3, 1983, and Jan. 24, April 5, and Sept. 26, 1984.)

———. 1987. *Computer Matching and Privacy Protection Act of 1987.* 100th Cong. 1st sess. 23 June 1987.

U.S. Congress. Office of Technology Assessment. September 1981. *Computer-Based National Information Systems.* Washington, D.C.: GPO.

———. June 1986. *Federal Government Information Technology: Electronic Record Systems and Individual Privacy.* Washington, D.C.: GPO.

———. April 1988. *Electronic Delivery of Public Assistance Benefits: Technology Options and Policy Issues.* Washington, D.C.: GPO.

U.S. Congress. Senate. Committee on the Judiciary. Subcommittee on Administrative Practice and Procedure. 1967. *Computer Privacy Hearings.* 90th Cong. 1st sess. 14–15 March 1967.

_____. 1967a. *Government Dossier.* Survey of Information contained in Government Files. 90th Cong. 1st sess.

_____. 1967b. *Invasions of Privacy.* Government Agencies, Hearings, 89th Cong. 2nd sess. 23–30 March, 7–9, 14, 16 June 1966.

U.S. Congress. Senate. Committee on the Judiciary. Subcommittee on Constitutional Rights. 1971. *Federal Data Banks, Computers and the Bill of Rights.* Hearings. 92nd Cong. 1st sess. 24–25 Feb., 2–4, 9–11, 15, 17 March 1971.

U.S. Congress. Senate. Committee on Government Operations. Ad Hoc Subcommittee on Privacy and Information Systems, and Committee on the Judiciary, Subcommittee on Constitutional Rights. 1974. *Privacy—The Collection, Use and Computerization of Personal Data.* Joint Hearings. 93rd Cong. 2nd sess. 80–20 June 1974.

U.S. Congress. Senate. Committee on Governmental Affairs. Subcommittee on Oversight of Government Management. 1983. *Oversight of Computer Matching to Detect Fraud and Mismanagement in Government Programs.* 97th Cong. 2nd sess. 15–16 December 1982.

_____. 1984. *Computer Matching: Taxpayer Records.* Hearings. 98th Cong. 2nd sess.

_____. 16 September 1986. *Computer Matching and Privacy Protection Act.* Hearings. 99th Cong. 2nd sess.

U.S. Department of Health, Education and Welfare, Secretary's Advisory Committee on Automated Personal Data Systems. 1973. *Records, Computers and the Rights of Citizens.* Washington, D.C.: GPO.

Weiss, Laura B. 26 February 1983. "Government Steps Up Use of Computer Matching to Find Fraud in Programs." *Congressional Quarterly Weekly Report.*

Westin, Alan F. 1967. *Privacy and Freedom.* New York: Atheneum.

Westin, Alan F., and Michael A. Baker. 1972. *Data Banks in a Free Society.* New York: Quadrangel/The New York Times Book Co.

Winner, Langdon. 1977. *Autonomous Technology: Technics-out-of-Control as a Theme in Political Thought.* Cambridge: MIT Press.

10

Economic Policy, International Competitiveness, and the Role of Technology Policy

Gary C. Bryner

The declining competitiveness of U.S. companies in global markets has become one of the most important issues on the nation's policy agenda. Evidence of the declining position of American industries in global markets is widespread. U.S. industries that dominated world markets two decades ago have seen their share of global markets shrink dramatically during the past two decades. U.S. firms that developed the technology for video cassette recorders and color televisions now only have less than 2 percent of the VCR and 10 percent of the TV markets. Japanese sales of semiconductor chips, another product developed by American scientists, have eclipsed sales by U.S. companies; the U.S. share of world markets has fallen from 45 to 35 percent since 1985. If current relative rates of growth continue, Japan will replace the U.S. as the leading manufacturer of electronics in the world before the mid-1990s (U.S. Department of Commerce 1990; Gephardt 1990).

Other indicators of the competitive position of the U.S. economy in the world show similar dramatic decline. The United States, long the world's major creditor nation, has now become the largest debtor nation, with a net indebtedness rapidly approaching one trillion dollars. The balance of trade account position of the U.S. has shifted from a longstanding position of a net exporter to one of major trade deficits. The merchandise trade balance deficit reached more than $150 billion in 1987 and the 1990 deficit is projected to be at least $90 billion. In 1987 more than two-thirds of the deficit was with Japan and the other industrialized Asian nations. Imports increased at an average rate of eight percent a year throughout the 1980s, well ahead of export activity (U.S. Congress, Office of Technology Assessment 1988, 2–4; 1990, 3–4).

American companies have become increasingly dependent on foreign suppliers for many of the critical components of manufactured products. The emergence of market opportunities in Eastern Europe and in developing countries focus attention on how well U.S. industry is able to participate in the rapidly evolving global economy. The optimism of the 1950s and 1960s and the confidence in America's technological and industrial preeminence has been replaced with doubts about the role the U.S. will play in a global economy where U.S. firms have become overshadowed by Asian and European enterprises (Lawrence 1990).

The political and economic stakes of this debate are enormous. As superpower tensions have decreased during the late 1980s and early 1990s, international relations have increasingly emphasized economic security. Domestic politics continues to give primary attention to the health of the economy in general and to levels of employment, wages, and economic growth. The threat to U.S. economic power from Japan and a united Europe has increased each year. Recent Congresses and administrations have engaged in a broad ranging debate over the general health of the economy and the prospects for the competitiveness of U.S. industries. The competitiveness of U.S. firms in global markets has major implications for the standard of living Americans will enjoy in the future. World economic leadership also has important symbolic importance to Americans and help determine the influence the U.S. is able to exert in international affairs (Young 1988). The success of policy makers and industry leaders in promoting American competitiveness will determine the extent to which resources will be available to pursue other important purposes such as alleviating poverty and promoting public health and protecting environmental quality and preserving natural resources.

How well have Congress and other policy makers responded to the threat of declining competitiveness? What policies in place and how well suited are they to addressing these concerns? What current policies should be terminated and what additional policies should be pursued? Policy making for competitiveness, like many of the other kinds of policies discussed in the other chapters of this book, must respond to two major challenges. First, there are levels of uncertainty concerning the consequences of alternative decisions, long lead times and lag times before errors are likely to be discovered, and potentially high costs in the event errors are made. There is tremendous uncertainty surrounding decisions concerning which new technologies to subsidize or regulate. Investments made now will take years before benefits are realized, or failure is apparent, and remedial steps can be taken. A primary challenge is for policy makers to learn from experience, to be able to make adjustments in a dynamic environment.

The second challenge for policy makers is that public policy making is only part of the decision making that affects competitiveness. Congress and other government institutions are limited in terms of what they can accomplish with the incentives and sanctions at their disposal. Decisions made independent of them by industry officials will play a major, probably a determining role in those decisions that most affect competitiveness. The key challenge for government policy makers is to serve as catalysts— to encourage private sector actions that are in harmony with public purposes. The purpose of this chapter is to examine how well Congress has been able to respond to these two challenges that are at the heart of policy making affecting the competitive position of U.S. companies in global markets.

The Importance of High-Technology Industries

While the policy debate over economic policy and international trade is concerned with the health and performance of American industry in general, much of the attention has been aimed at industries that rely heavily on continual technological innovation, particularly the high technology industries such as computers, telecommunications, and consumer electronics. In some economic sectors such as agriculture and aerospace, U.S. firms enjoy a strong position in global markets. In other sectors, where American companies no longer dominate international trade, such as steel and automobiles, the key to future competitiveness lies in industry's ability to harness technological innovation in competing with Asian and European producers. Nothing has demonstrated more clearly the importance of science and technology in economic activity than the decline of the U.S. consumer electronics industry. The dominance enjoyed by that industry in the 1950s and 60s has evaporated, and efforts to prevent further erosion will require a major commitment to the rapid development and application of new technologies to the design and production of electronic products.

The electronics industry deserves particular attention because of the primary position it plays in the U.S. economy and in international trade. The electronics sector exports one-fourth of its output, and the value of these exports have increased by an average of 18 percent a year since the mid-1970s. More jobs have been created during the last decade in the electronics industry than in any other, and it currently employs almost ten percent of the manufacturing work force. Electronics industries are the leading force for technological innovation in the U.S.; more patents are received by these firms than any other. Products produced by electronics firms have become critical elements in a wide range of industries as manufacturers increasingly rely on sophisticated computers, instrumentation, and communications systems. Electronics technology, particularly communications, is a central element of the growing service sector. National security is heavily dependent on a strong electronics capability (U.S. Department of Commerce 1990). The electronics and telecommunications industries converge in information technology, a central element to all aspects of modern economic activity.

The U.S. electronics industry is still the largest in the world, but virtually all indicators show a significant weakening in its global position. Since 1984, the U.S. has had the lowest growth rate in production. Japan has replaced the U.S. as the leading exporter of electronic goods. The electronics work force in the U.S. is stable while that of other countries is growing rapidly. American firms received 80 percent of patents for electronics in the U.S. in 1974, but that number declined to 55 percent in the late 1980s (U.S. Department of Commerce 1990).

Causes of the Weakening in American Competitiveness

There is no consensus over how the challenge of global economic competitiveness should be defined and what are its causes. Many economists argue that competitiveness is largely a function of exchange and interest rates and other macroeconomic variables, while others focus on the role of government trade policies such as subsidies and protection to domestic industries. Some emphasize balance of trade indicators, while others focus on the market share of U.S. firms. Competitiveness may be seen primarily as a question of the adequacy of infrastructure and investment and the position that U.S. firms are placing themselves for future global economic activity, or in terms of how U.S., Japanese, and Western European industrial activity compare in areas of research and development, costs of capital, and industry-government interactions. For others, competitiveness is ultimately a question of economic growth and the consequences for the standard of living enjoyed by Americans (Porter 1990). Explanations for the declining competitiveness of American industry range from the structure and operation of firms to general economic conditions. The most prominent are briefly summarized below.

Managerial and Worker Problems

Increased worker unrest and employee sabotage, a reduced commitment to the work ethic, and poor education skills and preparation have been blamed for declining productivity. Some place the blame on managers who are obsessed with short-run profits at the expense of investments that would produce long-run benefits and on paper entrepreneurship—making money through buying and selling companies rather than producing new or better products. Others emphasize the failure to recruit the best possible employees or the failure to move quickly to market new products (Magaziner and Reich 1982).

Reliance on Mass Production

Once the engine that drove the American economic engine, mass production is becoming obsolete in some industrial sectors. Global customers have increasingly come to demand high-quality products that are tailored to their specific needs and can be produced rapidly. Companies in other countries have become more adept than their U.S. counterparts in responding to these kinds of demands. American companies have fallen behind in their ability to improve product design and integrate it more effectively with manufacturing technologies (Dertouzos, Lester, and Solow 1989).

Government Regulation

The explosion in regulatory legislation and programs created in the early 1970s have been blamed for driving up the costs of products manufactured in the U.S. Particularly criticized were environmental regulations that were among the most stringent in the world. One of the first actions taken by the Reagan administration when it took office in 1981 was to create a regulatory relief task force headed by then Vice-President Bush to reduce the regulatory burdens on industry. The task force and other changes in the regulatory process, including review of all proposed and final major regulations by the Office of Management and Budget, resulted in a significant drop in the number of regulations issued by regulatory agencies and, arguably, a reduction in the stringency of many new regulations (U.S. Office of Management and Budget 1985–88). Antitrust regulations are also believed by many to prohibit research and other cooperative efforts among industries that would foster their competitive position in global markets (Browning 1991).

Defense Spending

Both defense procurement and research and development spending have traditionally been viewed as producing spinoffs that benefit civilian industries. While there are a number of such positive spinoffs, such as the development of a huge military transport plane that produced Boeing's 747 jumbo jet, critics have argued that defense research and defense spending is an overall drag on civilian industry. Defense needs that drive R&D do not always yield trickle down benefits to commercial industry, and U.S. industries have lost out to Japanese firms that have had civilian markets in mind from the earliest stages of development. Cost control, fast market response, and other characteristics of successful competition in civilian markets are not exactly fostered in military R&D and procurement. While defense spending has had some spillover benefits, the real question is whether industrial competitiveness would be enhanced by more direct federal intervention than by relying primarily on the occasional and imperfect spinoffs from defense spending (Brooks and Branscomb 1989).

The Adversarial Nature of American Business and Society

The rise of product liability litigation is often cited by business executives as one of the most serious constraints they face. Manufacturing and insurance companies complain that liability awards to plaintiffs in defective product suit are excessively expensive and add significantly to the costs of doing business and dampen innovation and experimentation in industry (Huber 1988).

Increased Competition from Abroad

Foreign companies that provided little competition to American manufacturers in the 1950s and 60s have blossomed. They rely on plants and equipment that are in some cases much newer than used in American companies. They benefitted from U.S. aid after World War II and the continual dissemination of information and technology. Cooperative institutions in Japan have fostered coordination and integration of efforts among high-technology industries (Nanto and McLoughlin 1991). While the U.S. has played the leading role in basic science, other nations have concentrated on commercial application of basic research done elsewhere. Countries such as Japan and Germany spend comparatively little on defense research and development and procurement, freeing resources for consumer products. Some blame is also directed to the actions of foreign governments in protecting their domestic industries by limiting imports from other countries and subsidizing exports. U.S. industries are disadvantaged by such practices that make it difficult to export U.S. goods in these countries and give advantages to imported goods in U.S. markets. Much of the blame for the imbalance of trade between the U.S. and Japan, from this perspective, focuses on Japanese government trade policies that have no countervailing American policies.

The High Cost of Capital

High interest rates during the last decade have raised the price of investments for improved productivity and output. High interest rates, in turn, are a result of Federal Reserve System efforts to check inflationary pressures, reduced savings on the part of Americans, high budget deficits and the demand for government borrowing, and other characteristics of the American economy that affect investment and innovation in American firms.

Chronic Economic Problems

Stagnation is believed to be a chronic, inherent element of capitalism by Marxists and other critics of the U.S. economic system. As large corporations dominate markets, they have insufficient opportunities for profitable investments. This eventually results in inadequate demand for plants, machinery, raw materials, and labor. Unemployment and reduced demand for consumer goods complete the cycle. U.S. production peaked in the early 1970s as global markets expanded; the lack of continued growth in demand has resulted in excess productive capacity. Investments and other steps to stimulate productivity will only worsen the problem. In contrast to these other explanations, the solution here lies in efforts to stimulate demand such as redistributing global wealth (Schlefer 1989).

Formulating Policy: How to Promote High-technology Industries?

There are numerous possible policy responses to the problem of the decline of American industrial competitiveness, mirroring the alternative explanations for the problems of competitiveness. Since there is no agreement over exactly how to define national competitiveness, there is little consensus over what policies should be pursued. Much of the debate in Congress and elsewhere over the role of government in addressing the challenge of global competition has focused on finding an overall policy approach that is located somewhere between the extremes of laissez-faire domestic policies and unilateral free trade and an industrial policy of promoting specific domestic industries and carefully regulating imports and exports. Most of the debate has focused on macroeconomic policies to reduce interest rates and ensure a healthy economy, increased government spending for basic science and incentives for more R&D in industry, increased spending and incentives for commercialization of new technologies, reduced regulatory burdens, and assistance to export-oriented industries and to domestic producers.

Most options for improving the competitive position of U.S. industries can be grouped into one of four sets of policy possibilities. A first category of options is based on the view that policies should be aimed at promoting a strong, growing economy. The primary concern here is reducing the cost of capital by cutting interest rates, capital gains and other taxes, the federal budget deficit, and by increasing saving rates. Are these efforts sufficient or is more direct intervention required? Are these steps incompatible with competing policy concerns such as spending on social programs or national defense and maintain a progressive tax structure? An analysis of such a wide-ranging set of policies is beyond the scope of this chapter. It is clear that many people inside and outside of Congress believe that macroeconomic policies are not enough and that they must be combined with more direct efforts to promote competitiveness.

A second set of policies affect the competitive position of industry by imposing regulatory requirements. Some regulations, such as environmental, occupational health and safety, and consumer protection standards, add to the costs of production and distribution. Antitrust regulation limits the extent to which industries can join forces in common research and development efforts. Liability laws impose financial responsibility on producers for injuries resulting form the use of their products. Restrictions have also placed on the production of certain kinds of products resulting from research in biotechnology. Should environmental, occupational, and consumer regulations be reduced in order to facilitate competitiveness? Should joint R&D efforts be treated differently than other restrictions on industry cooperative efforts? Can export restrictions be eased now that the communist world is in decline? Is there sufficient competition in global markets to reduce concerns about concentration of market power and price-fixing? Answers to many of these questions require examination of a wide range of concerns regarding regulation that are beyond the scope of this paper. While many policy analysts have championed cooperative R&D efforts and concluded that the focus of anti-trust policy might best be directed to ensure global rather than local competition, others have argued that local competition is the key to global performance (Porter 1990b). There is likely to be relatively little political support for reduced environmental, health, and safety regulations in order to improve competitiveness as public opinion polls regularly report strong support for these regulatory programs (Mitchell 1990).

A third set of possible policy interventions would encourage improvements in industry management practices. Excessive industry concern with short-run profits at the expense of long-run productivity, obsession with "paper entrepreneurship"—takeovers and litigation rather than actual production of goods and services—and inflexibility in production methods have been blamed for much of the decline of American competitiveness. While this area may be least susceptible to government remedies, tax and other policies might be fashioned to discourage certain kinds of behavior and encourage others. Should efforts by foreign companies to purchase and operate businesses in the U.S. be encouraged as a way to stimulate changes in outmoded industrial practices? What risks are involved in efforts by government to "improve" corporate management practices?

Answers to many of these questions also go well beyond the scope of this chapter and very likely beyond the scope of public policy in confronting issues that have traditionally been viewed as private managerial prerogatives.

The fourth set of policy options involve government subsidies to promote research and development and commercial application. Some subsidies are aimed at promoting basic scientific research or education levels, while others are aimed at development and application of technologies to commercial or military purposes. Subsidies can take the form of grants to engineering and science students and teachers, funding of demonstration projects, government-industry joint venture efforts, and loan guarantees and other direct assistance to specific firms. Subsidies may take a more indirect form, such as patent and copyright protections, or import quotas and tariffs. These policies have engendered a host of questions: Should the federal government subsidize "big" scientific projects, such as the supercollider or space station, or fund a wider range of efforts? What is the interaction of defense and non-defense research and development spending? To what extent are basic elements of scientific research such as peer review at odds with the need to keep information secret in preparation for patenting and other steps involved in the commercialization of research? Should government subsidies be limited to basic science? Should they be extended to promote technological and commercial applications of basic research? Should particular industries or even firms be singled out for special help?

These questions get at the heart of how competitiveness policy might best be understood and what kinds of policies can stimulate the productivity of American industry. Increases in productivity that are sustained over the long run are the key to economic growth and improved standard of living since they result in real increases in output. Productivity is a function of the general health of the economy and the societal infrastructure on which it rests as well as what goes on in specific sectors of the economy. Not all industries need be or probably can be dominant in world markets. The key to competitiveness is how well policy makers can foster the competitive advantages enjoyed by U.S. industries and strengthen industries that are essential for other national interests.

There are a number of challenges confronting Congress and the executive branch in making policy to promote U.S. industrial competitiveness:

> If Congress is to mandate intervention, how can it be done in a way that will not simply help political constituents and protect declining industries?

How can Congress promote more of a commitment to long-term growth and investment in industry when members themselves seem to be primarily responsive to short-run, political incentives?

How can Congress deal with the uncertainty of knowing which technologies are most promising? How can policy making best take into account the uncertainty, lag time between policy intervention and results, and other characteristics of competitiveness-related policies?

Should Congress fund basic research, and concentrate on enlarging the pool of knowledge, or should it emphasize development and application?

How can the strengths of small-scale scientific research be fostered while also providing funds for the large-scale projects?

What should be the relationship between military and civilian R&D?

How can Congress and the president organize the actions of federal agencies to ensure effective and efficient policy implementation?

How can Congress make effective policy when so many determinants of success are in the hands of non-government actors?

Current Policies in Place to Promote Competitiveness

Since the competitive position of American industry is, at least to some extent, a function of overall economic conditions, virtually every agency and program that regulates and promotes economic activity and general economic conditions plays a role in fashioning competitiveness policy. This is also true in Congress, where most congressional committees have at least some jurisdiction over economic and trade-related policies. Senate committees and subcommittees that are particularly involved in the promotion of high-technology industry include:

- Appropriations;
- Commerce, Science, and Transportation;
    Communications, Foreign Commerce and Tourism, and Science, Technology, and Space;
- Energy and Natural Resources;
- Finance;
    International Trade; and
- Judiciary;
    Antitrust, Monopolies, and Business Rights.

House Committees and Subcommittees include:

- Appropriations;
- Energy and Commerce;

Telecommunications and Finance, Energy and Power, Commerce, Consumer Protection, and Competitiveness;
- Judiciary;
  Science, Space, and Technology, Energy Research and Development, Science, Research and Technology, International Science Cooperation; and
- Ways and Means;
  Trade.

The Joint Economic Committee's Economic Growth, Trade, and Taxes; Resources and Competitiveness; and Investment Jobs, and Prices subcommittees also play a major role.

The Bush administration has opposed any effort that smacks of industrial policy—"the notion that the government should be picking which industries and technologies are winners"—and argues for policy actions that are limited to providing a healthy economic environment for growth (Boskin 1990). The president's key economic advisers champion free market approaches and overall growth in the economy as the keys to competitiveness. Nevertheless, the White House and the executive branch have several administrative means for shaping competitiveness policy. The major agencies in the Executive Office of the President that deal with trade and technology issues are the Office of Special Trade Representative, the lead agency for negotiating bilateral and multilateral trade agreements; the Office of Science and Technology Policy, responsible for advising the president on policy issues concerning science and technology; the Council of Economic Advisers, also an advisory body to the president; and the Office of Management and Budget, the critical agency here because of its control over federal agency budgets and its review authority over agency regulatory and legislative initiatives.

Despite White House rhetoric, there are a great number of policy efforts that affect competitiveness. Several agencies have been created to promote the export of U.S. products. The Department of Commerce advertises itself as a ready ally of U.S. exporters, providing offices in sixty-eight American cities to help businesses seeking to sell their products abroad. Several bureaus—the Economic Development Administration, the Bureau of Export Administration, the International Trade Administration, the National Institute of Standards and Technology, and the Patent and Trademark Office play critical roles in boosting U.S. industrial activity. The Foreign Agriculture Service in the Department of Agriculture encourages agricultural exports by providing marketing information, export credit programs, and food assistance programs to foreign nations. The Export-Import Bank offers special financing programs and offers assistance to companies facing foreign government-subsidized competition. The Small Business Administrations helps smaller companies locate markets and meet regulatory and other legal requirements of international trade. The Overseas Private Investment Corporation (OPIC) facilitates private investment efforts in developing countries (U.S. Department of Commerce 1990). Research funded by the National Institutes of Health have been enormously important in the development of the biotech industry (Carey 1990, 48). A host of regulatory agencies impose standards on American industries, including the Environmental Protection Agency, the Occupational Safety and Health Administration, and the Securities and Exchange Commission. The Federal Trade Commission and the Justice Department share responsibility for implementing anti-trust laws. The most important policy effort to promote U.S. high-tech industries is conducted by the Defense Department. The Pentagon's Defense Advanced Research Projects Agency (DARPA), for example, funds military research that has had numerous civilian applications (Congressional Quarterly 1988).

There are several different ways in which the Federal government sponsors research and development that ultimately affects the competitiveness of American industry. These efforts can very roughly be plotted on a continuum with one end basic research and infrastructure-building, and the promotion of commercial application at the other end. The Federal budget has traditionally included funds for research and development in areas of basic science where private companies are not working because of high costs and risks or the lack of direct commercial application. Federal agencies fund a variety of efforts to develop an infrastructure of transportation facilities, an educated work force, research institutes, and other resources that industrial activity requires. Applied research has primarily been funded by agencies that are the primary user of the results in pursuing their missions, or where national security interests are advanced through the development of new technologies. In all of these cases, there may be commercial benefits or spinoffs from these Federal efforts, but their primary focus is the pursuit of other purposes—American leadership in science, providing an effective base for economic activity, and national security. However, recent legislation requires federal agencies to foster the commercialization of the results of research and development conducted for them by making technology transfer part of their missions, establishing cooperative research between Federal laboratories and industry, and giving particular attention to assisting small businesses (U.S. Department of Commerce 1990; Gephardt 1990).

Some Federal efforts have moved considerably along the continuum toward funding programs that stimulate innovation and encourage research, development, commercialization, and diffusion of new technologies. Some policies seek to facilitate private sector cooperation; the National Cooperative Research Act encourages companies to undertake joint research by limiting the application of anti-trust laws to such ventures (P.L. 98–462). The Economic Recovery Act of 1981 (P.L. 97–34), the Tax Reform Act of 1986 (P.L. 99–514), and the Omnibus Reconciliation Act of 1989 (P.L. 101–239) provided tax credits for payments by companies to universities for R&D work and tax deductions for contributions of equipment made to universities. Amendments to patent and trademark laws give to universities, small businesses, and non-profit organizations title to inventions produced with federal funds, creating new incentives for these institutions to develop commercial applications of their R&D activities.

Even more directly aimed at the eventual commercialization of new technologies have been cooperative efforts among industries and between industries and universities organized by Federal agencies. Engineering Research Centers, Science and Technology Centers, and Industry-University Cooperative Research Programs have been established by the National Science Foundation. These programs have been broadly directed at all industries and are generally aimed at research efforts rather than the development of specific technologies. Recent initiatives, have gone even further to assist specific industries that are believed to be critical to the larger industrial sector of the economy. Joint R&D efforts between Federal laboratories and industry and grants in specific industries such as semiconductor manufacturing, superconductivity, and steel have been promoted as essential in pursuit of national defense and economic security. The great variety in Federal R&D programs are perhaps best explored in more detail by examining related provisions of the most recent Federal budget.

The Federal Budget for R&D and Physical and Human Capital

Federal spending for research and development in 1990 was projected to reach more than $62 billion. Total R&D spending in the U.S. was expected to reach $132 billion, with the Federal government's share about 46 percent of the total. The Federal share of R&D spending has declined considerably: twenty five years, ago, the Federal share was approximately two-thirds of all R&D spending. The amount of money spent for R&D in the U.S. is greater than total spent in Japan, Germany, France, and the United Kingdom only because of the size of the U.S. economy. These countries all spend about 2 to 2 1/2 percent of their GNPs on R&D. One major difference between spending in the U.S. and its competitors is the share provided by the Federal Government (higher in the U.S.) and the percentage of R&D spending that goes to defense (65 percent). Other major areas of Federal R&D include health (11 percent) and space (6 percent) (U.S. Office of Management and Budget 1990, 28–29). Most R&D

TABLE 10.1   Summary of Federal Investment

	1960	1970	1980	1985	1990
Total Investment[a]	$31.4	$56.5	$119.4	$184.8	$231.8
R&D	7.3	15.2	30.2	47.2	62.2
Human capital	1.7	8.2	26.2	23.0	28.2
Physical capital	22.4	33.2	63.0	114.6	131.6
Total Investment[b]	110.4	148.0	140.9	169.0	185.9
R&D	33.2	44.4	36.5	42.4	47.1
Human capital	7.4	22.4	31.1	20.0	20.9
Physical capital	69.8	81.3	73.3	106.6	118.0
Total investment[c]	34.1	28.9	20.2	19.5	19.4
R&D	7.9	7.7	5.1	5.0	5.2
Human capital	1.8	4.2	4.4	2.4	2.4
Physical capital	24.3	17.0	10.7	12.1	11.0

[a] In billions of $.
[b] In billions of 1982 $.
[c] As a percentage of outlays.

Source: Executive Office of the President, Office of Management and Budget (1990), *Budget of the United States Government, Fiscal Year 1991* (Washington, D.C.: GPO), p. 39.

is performed by industry (72 percent); 11 percent takes place at colleges and universities, 11 percent is conducted by Federal agencies, and six percent at nonprofit organizations, state and local government agencies, and Federally funded R&D centers (70–71). Another major difference is the share of R&D spending on industrial development of new technologies, ranging from 0.2% in the U.S. to 4.8% in Japan to a whopping 15.3% in Germany (Rubin 1991). Table 10.1 shows the trend in R&D spending in the Federal budget. Spending declined throughout the 1970s and early 1980s; by the late 1980s, however, it approached the 1970 level of spending.

There is a significant difference between defense and nondefense R&D in terms of the allocation to basic research, applied, and actual product development. Defense R&D spending is much more oriented toward application than nondefense spending. Basic research comprises about 16 percent of total Federal Government R&D spending. In defense, about 90 percent is for development, 7 percent for applied research, and 3 percent for basic research. In nondefense R&D spending, in contrast, about 29 percent goes to development, 29 percent to applied, and 42 percent to basic research (Carey 1990, 47).

Research and Development

Funding for basic research by the Federal government is dominated by health research, which accounts for nearly half of all spending for basic research, followed by energy, space exploration, and national defense (U.S. Office of Management and Budget 1990, 68). National Science Foundation outlays extend to wide range of concerns. The fiscal year 1991 budget included major increases in funding for the National Science Foundation and for the following specific R&D areas as outlined in Table 10.2. The table also demonstrates the much higher level of funding given to applied rather than basic research, and the dominance of defense-related research and development in applied areas.

Many programs are aimed at the developing new technologies that will assist Federal agencies in pursuing their own missions and supporting "generic" R&D rather than aimed at direct commercial applications defined as R&D that has a "broad national benefit:" "pre-competitive, so that its benefits cannot be fully or quickly captured by specific companies;" and involves a "significant level of participation and/or cost sharing with private sector partners commensurate with the expected benefits" (U.S. O.M.B. 1990, 84).

The Department of Commerce's National Institute of Standards and Technology has two programs designed to develop and transfer a variety of advanced technologies to the private sector. Six federal agencies, primarily NASA, fund research in robotics. Nearly $470 million is budgeted in 1991 for high performance computing, including software and networks, primarily at NASA and the Defense Advanced Research Projects Agency (DARPA). The largest program is a $100 million grant from DARPA to SEMATECH, a R&D consortium established by the semiconductor industry. DARPA funds about one-half of the consortium's effort to improve semiconductor manufacturing techniques. DARPA and other agencies have also funded research into advanced imaging technologies, including interactive graphics, high definition displays, and, until recently, High Definition TV (U.S. O.M.B. 1990, 85–6).

Research in three areas has been given particular support by the Federal government: space exploration research, biotechnology, and the Superconducting Super Collider. Federal government sponsorship of a space exploration program has traditionally been defended as producing a wide range of commercial spinoffs in addition to its military and scientific value. One-half of the nearly $16 billion for civilian space programs is aimed at building a transportation infrastructure, primarily expendable launch vehicles and the Space Shuttle. The Moon/Mars manned space initiative emphasizes research in robotics, high-resolution image processing, bio-

TABLE 10.2 Research and Development in the FY 1991 and 1992 Federal Budgets (budget authority, $ millions)

	1990 Enacted	1991 Enacted	1992 Proposed	1990–1992 % Change
Major initiatives[a]				
Doubling the NSF budget	2,084	2,316	2,722	+ 31
Global climate change research	659	954	1,186	+ 80
Agricultural research initiative	43	73	125	+ 290
HIV/AIDS	1,162	1,152	1,210	+ 4
Space station freedom	1,928	2,044	2,214	+ 15
Magnetic levitation transportation	2	12	24	+1,100
Superconducting super collider	228	243	534	+ 134
High performance computing	448	489	638	+ 42
Natural resource research	710	844	900	+ 27
Military research and development	39,877	37,783	43,247	+ 8
Government-wide Totals				
Conduct of R&D				
Basic research	11,398	12,320	13,320	+ 17
Civilian	10,459	11,296	12,278	+ 17
Defense[b]	939	1,024	1,041	+ 11
Applied research and development	52,313	51,791	58,758	+ 12
Civilian	13,375	15,031	16,552	+ 24
Defense[b]	38,938	36,760	42,206	+ 8
Subtotal: Conduct of R&D	63,711	64,111	72,078	+ 13
R&D Facilities	3,023	3,082	3,545	+ 17
Total	66,734	67,192	75,623	+ 13

Addends may not add to totals because of rounding.

[a]Major initiatives include funds for research and development that also are included in the government-wide totals for R&D as well as funds from other non-R&D programs.

[b]Includes military-related programs of the Departments of Defense and Energy.

Source: Executive Office of the President, Office of Management and Budget (1990), *Budget of the United States Government, Fiscal Year 1991* (Washington D.C.: GPO, p. 68); Executive Office of the President, Office of Management and Budget (1991) *Budget of the United States Government, Fiscal Year 1992* (Washington D.C.: GPO), pp. 35–87.

logical and space biomedicine, life support systems, and high energy power sources. As indicated above, a sizeable component of space research is aimed at increasing understanding of global climate change and ozone depletion through satellite monitoring of oceans, droughts, the chemistry of the upper atmosphere, and other environment-related phenomena. The commercial space sector is primarily communications satellites, but is expanding into other ways of developing commercial opportunities in space through the creation by NASA of Centers for the Commercial Development of Space, cooperative programs between universities and companies (U.S. O.M.B. 1990, 55–8).

Federal support of research in biotechnology has been championed as a means of improving the quality and quantity of the food supply and preventing or developing treatments, for disease. Twelve Federal agencies, including the Departments of Health and Human Services, Energy, Commerce, Defense, and Agriculture, and NASA and NSF fund more than $3 billion dollars of research each year. The National Institute of Health-sponsored basic research includes the Human Genome Initiative, a long term project to identify and map each of the estimated 10,000 human genes and then determine the genetic causes of diseases and develop genetic-based therapies. NIH has also entered into more 200 cooperative R&D agreements with private firms. Grants to universities and Federal agricultural research laboratories are aimed at increasing agricultural output, improving the safety of the food supply, and minimizing environmental damage from the use of pesticides. More than 400 biotechnology firms have been created in the last 15 years; another 200 firms have diversified into this area; and 200 more provide supporting equipment, materials, and services. The tremendous commercial potential for the prevention and treatment of cancer and other diseases has resulted in major research initiatives in drug companies (U.S. O.M.B. 1990, 59–62; Waldholz 1990). High energy physics research seeks to identify and understand the behavior of the universe's most elementary particles. These efforts require accelerators that cause particles to travel at velocities approaching the speed of light. The Department of Energy and the NSF will fund nearly a billion dollars worth of high energy physics research in 1990. The Federal government, the state of Texas, and private industries have agreed to join in funding the construction of the Superconducting Super Collider that will accelerate particles some 20 times faster than existing facilities in a 53 mile long circular tunnel. The SSC is promoted as a symbol of American leadership in science and technology and a key to increasing basic scientific knowledge that will eventually have commercial spinoffs. The 1991 budget calls for more than $300 million for the SSC; the project was estimated in 1986 to cost $4.4 billion but the latest projections range from $7.8 to $11.7 billion (U.S. O.M.B. 1990, 63–5).

Human Capital and Investments in Education

Government R&D expenditures can also be compared with two other categories of investment—human and physical capital—as indicated in Table 10.1. Government spending for education and training can, of course, only partly be understood as investments in human capital as they have much broader purposes and benefits. The skills and training of the nation's workforce, however, play an important role in determining productivity. Human capital investments by the Federal Government have increased tremendously since 1960. By 1980, levels of investments in human capital and R&D were almost equivalent. However, shifts in budget priorities away from social programs in the 1980s and toward increased defense outlays resulted in R&D spending more than doubling outlays for human capital since then. Declining performance on standardized tests; increase crime, drug use, and related problems in schools; declining interest in science studies and the increase in the percentage of science doctoral degrees going to foreign students at U.S. universities; lack of basic skills and other prerequisites for effective on-the-job performance by graduates of public schools; and other problems have been widely discussed as contributing to the competitiveness problem (National Commission on Excellence in Education 1983; Johnston 1985; Chubb and Moe 1990).

Five Federal agencies sponsor programs to improve science, mathematics, and engineering education. Funding for these programs is expected to grow from $849 million in 1990 to $1.061 billion in 1991. Some 45 percent of the funds go to the National Science Foundation for workshops for teachers and the development of innovative teaching materials, research opportunities in science laboratories for high school and college students, and innovative programs at science museums. The second largest program ($287 million in 1990) is the National Institutes of Health, where almost 12,000 graduate trainees work in research laboratories in programs and thousands of other students are supported through grants given to faculty members (U.S. O.M.B. 1990, 87–9).

Physical Capital

Federal government investment in physical capital serves a variety of public purposes in addition to directly enhancing competitiveness. National defense accounts for about two-thirds of these expenditures (U.S. O.M.B. 1990, 34). Other major investments include space programs, the air traffic control system, federal agency purchase of computers, and Army Corps of Engineers projects. Most of the spending for competitiveness-related investments take the form of grants to state and local governments. More than half of these grants are for highways; the balance goes to regional and community development programs, mass transit, pollution

TABLE 10.3   Federal Investment in Physical Capital ($ million)

	1960	1970	1980	1985	est. 1990
Direct investment[a]					
Defense	$17.2	$23.6	$32.5	$78.0	$90.4
Nondefense	1.9	2.5	8.1	11.7	15.5
Subtotal	19.1	26.1	40.5	89.7	105.9
Grants to state and local governments[a]	3.3	7.1	22.5	24.9	25.3
Total	22.4	33.2	63.0	114.6	131.2
Direct investment[b]					
Defense	51.7	55.8	39.6	72.3	84.0
Nondefense	5.9	6.1	9.1	11.4	14.1
Subtotal	57.6	61.9	48.7	83.7	98.1
Grants to state and local governments[b]	12.2	19.4	24.6	22.9	19.9
Total	69.8	81.3	73.3	106.6	118.0
Direct investment[c]					
Defense	18.6	12.1	5.5	8.2	7.5
Nondefense	2.1	1.3	1.4	1.2	1.3
Subtotal	20.7	13.4	6.9	9.5	8.8
Grants to state and local governments[c]	3.6	3.6	3.8	2.6	2.1
Total	24.3	17.0	10.7	12.1	11.0

[a]Current $.
[b]Constant 1982 $.
[c]Percentage of total outlays, current $.

Source: Executive Office of the President. Office of Management and Budget (1990) *Budget of the United States Government, Fiscal Year 1991* (Washington, D.C.: GPO) p. 36.

control facilities, and airports. These grants expanded throughout the 1950s, 60s, and 70s, particularly as the interstate highway system was built, and then declined in the 1980s as budget cuts were concentrated in these and other areas (U.S. O.M.B. 1990, 34–5). Table 10.3 summarizes the trend in a dramatic decline in Federal financing of physical capital.

Transportation is a particularly important element of the underlying infrastructure of the economy (U.S. O.M.B. 1990, A-929). The Department of Transportation proposed a number of initiatives in the summer of 1990 to replace aging transportation infrastructure and engage the federal and state governments in finding ways of addressing transportation-related environmental problems, but the Office of Management and Budget eliminated from the plan the commitment of new resources to finance these programs (Berger 1990).

Energy is another critical component of industrial activity. The 1991 budget proposed a 30 percent increase from 1990 requested levels for research and development programs for solar and renewable energy and energy conservation. More than $43 million was proposed for solar photovoltaics research; $28 million for biofuels—particularly additional research for ethanol fuels and the incineration of energy from municipal waste—and $128 million for conservation—primarily funds to install more energy efficient lighting in federal facilities and to encourage the development of building conservation standards (U.S. O.M.B. 1990, 131).

Federal agencies help fund the construction and purchase of equipment and facilities at universities, laboratories, and research support facilities. The construction of R&D facilities at universities, in terms of real dollars, peaked in 1965, and concern with the deterioration of these facilities has resulted in the increasing use of earmarked appropriations by influential members of Congress. Between 1980 and 1989, more than $900 million was earmarked for over 300 specific projects, primarily located at universities. Funds for these programs do not undergo the same review that programs funded through a competitive process receive, and these funds are generally taken from other programs rather than representing additional spending. Most of these grants go to energy, defense, and space-related programs.

The Debate over Competitiveness Policies:
Policy Implementation and Evaluation

How well have Congress and the president promoted competitiveness? Should these funding decisions continue into the future? Should priorities for government intervention be altered? Are structural or administrative changes needed? Since there is no agreement in Congress over how define the competitiveness problem and what are its causes, it is difficult to assess the appropriateness of the existing set of Federal policies. Competitiveness is a function of a great number of variables, and policy efforts are only a part of a complex economic and political context. Policies to promote competitiveness may make major contributions to that goal, but nevertheless be overwhelmed by other developments. The results of policies aimed at fostering competitiveness may be so indirect that their impact on specific products cannot be determined. There is considerable time lag between the initiation of policy initiatives and when they bear fruit. Nevertheless, the discussion above points to a number of implications for the future of policy making for competitiveness.

First, there are a great number of federal and state programs aimed at fostering scientific research and technological development. States created more than 250 programs to encourage the development and commercialization of new technologies (Osborne 1990). There are numerous regulatory bodies, each with their own set of restrictions with which companies must comply, that complicate considerably the conduct of business. Congress is as fragmented in this policy area as it is elsewhere, and there is little coordination in congressional decision making and budgeting. Budget documents and processes provide little opportunity to track various decisions made concerning science and technology and assess the effectiveness of the spending programs (Brandon 1990). Despite efforts to coordinate competitiveness policy in the Executive Office of the President, there is little information about how effective and compatible these policy interventions are. Congress and the executive branch have not done enough in oversight efforts aimed at identifying and eliminating conflicting requirements and programs. Federal, not industry officials should be responsible for refereeing among competing requirements.

Second, the areas where U.S. industries have been most competitive in global markets are sectors of the economy such as aerospace, agriculture, and biotechnology where government, universities, and industry have been working closely together. Federal efforts have not been limited to basic research, but have promoted the development of technologies. Despite claims to the contrary, the Federal government has regularly been involved in the promotion of technological innovations with direct commercial applicability. Similarly, the countries that represent the greatest competitive challenge to the U.S. aggressively employ cooperative business-government efforts to pursue overall economic goals (Lawrence 1990).

Competitiveness-oriented policies are generally most effective when they move beyond traditional distributive politics and serve as catalysts to stimulate private sector behavior, rather than government bearing primary responsibility for providing services (Osborne 1990). Programs that provide matching start-up grants, disseminate information, and ensure more business involvement in university and other research centers can help harness the efforts of a greater number of actors than direct government provision of assistance can ever do. Policies that are based on clear outcomes and include measures to ensure regular policy evaluation are much more likely to enhance competitiveness than programs that are permitted to degenerate into assistance that benefits recipients without pressuring them to improve. Aggressive, catalyst policies are not as popular as government grants with no strings attached, and a major challenge for Congress is to push companies in ways that they may not want to go. Policies that assist existing industries to become more innovative are also as important as those aimed at emerging ones.

Perhaps most importantly, policies that reflect the dynamic nature of global economic activity are more effective than those that remain unchanged year after year. The role of government shifts as sectors of the economy evolve; incentives to encourage the development of new technologies will have to be replaced eventually with efforts to expand market shares, for example. Policies must go beyond providing basic infrastructure to encourage continual, sustained investment in advanced, specialized resources that build on the existing advantages enjoyed by the nation and to encourage the development of compatible and supporting industries (Porter 1990b).

Third, the commercial, civilian spinoffs from defense R&D are modest. The problems with defense R&D and procurement resulting in much more expensive and unreliable military hardware than is in the national interest have been widely documented (Fallows 1984; Stubbing 1986). Expectations for decreased defense spending and the wide gulf between military and civilian applications require that more direct efforts be undertaken to promote the development of technologies that are essential to overall economic, commercial, and industrial activity. U.S. firms no longer enjoy the luxury of taking years to translate basic scientific research into technological advances. As industrial activity relies more and more on electronic components, computers, and related technologies, the health of those basic technologies in the U.S. will become increasingly critical.

Fourth, a preoccupation with nationalistic promotion of industry in Congress and elsewhere is not in the national interest. It has become increasingly difficult to know which firms are essentially American and which are largely benefitting citizens of other nations. While assistance should be primarily aimed at firms located in the U.S., policy makers should recognize the advantages of global technological development. Increasing economic ties can help nations to recognize that their economic future is interdependent. Solutions to global problems of environmental degradation, poverty, disease, and malnutrition, like most everything else, are ultimately political issues, but technological progress will be an indispensable element if political solutions and cooperative efforts are to take place.

Fifth, the development of a transportation infrastructure has been of great importance in facilitating interstate commerce and facilitating trade between suppliers and manufacturers. As is true of Federal efforts to contribute to human capital, however, there is widespread criticism with the state of America's infrastructure and its implications for economic activity in general and competitiveness in particular. Nearly one-fifth of the interstate highway system has exceeded its intended life span. Thirteen percent of the system's bridges need repairing or replacing. More than $45 billion dollars (in 1989 dollars) is needed to repair and improve the nation's waterways and major ports. Another $10 billion is needed to upgrade airports, $30 billion to modernize and replace aging mass transit facilities, and more than $100 billion to expand water and sewer systems (Lamm, Caldwell, and Mehlman 1989; U.S. Department of Transportation 1990).

The hundreds of billions of dollars in spending that is required to improve bridges, dams, highways, airports, energy production facilities, and other infrastructures, are unlikely to be forthcoming in the near future given current budget deficits. Without increased spending for infrastructure, however, economic development in many areas of the nation will be impeded. One of the great challenges confronting Congress is how to go beyond dealing with the deficit to providing the kinds of investments that are needed to ensure a strong economy.

Sixth, insufficient attention has been given to investments in energy technologies and efficiency to help American industry meet the challenges that lie ahead in ensuring an adequate energy supply. The U.S. consumed in 1988 more than 80 quadrillion British thermal units (BTU's) of energy (one quadrillion BTU's or a "quad" is the energy supplied by about 500,000 barrels of oil a day for an entire year). About one third of that energy was used in transportation, another third was consumed by industry, and the balance by residential and commercial consumers. The amount of energy consumed has grown by about 8 percent since 1973. Economic output grew during that same period by 47 percent. Since the population was also growing at a modest rate during that period, per capita energy use declined on average by about one-half of one percent annually. The demand for electricity, however, actually increased at an average annual rate of 2.7 percent. The demand for transportation fuels also increased during these years by about 0.7 percent a year. Improved fuel efficiency was offset by a major increase in the number of vehicle miles driven (U.S. Department of Energy 1990).

The U.S. is the largest and one of the least efficient consumers of energy in the world. It uses one-fourth of the energy produced every year, but its economic output per unit of energy is less than one-half that of Japan and West Germany. In Japan, for example, energy consumption per $10,000 of gross domestic product was 2.9 metric tons of oil equivalent; in West Germany, 3.1 metric tons, Great Britain, 3.5, and Sweden, 4.0. The figure for the U.S. is 6.1 tons and 8.0 in Canada. Given the cost of energy, energy inefficient in the U.S. places industries in a real competitive disadvantage (World Resources Institute 1988, 114). Current methods of electrical production are rather inefficient. One estimate concluded that much of the energy produced in the U.S. was wasted in the production and distribution of electricity. Of the 28 quadrillion BTUs of electricity produced in the U.S. in 1987, only 8 "quads" were sold to consumers; the balance was used to generate and distribute the electricity. If fossil fuels were employed directly for heating far less fuel would need to be burned (Global Tomorrow Coalition, 1990, 196–97). State and federal policies can encourage the least cost production of electricity and more cost-effective means of heating homes and buildings.

One of the most promising areas for investment efforts that would contribute to improved competitiveness of American industry is in energy efficiency. While such efforts are easily justified on other grounds such as environmental protection, competitiveness makes the case for them even stronger. Tax incentives to encourage conservation-related investments, changes in government procurement, and elimination of counterproductive practices and subsidies promise economic and environmental benefits; they reduce costs and make U.S. products more competitive in international markets as well as reduce greenhouse gas emissions. They will put U.S. industries in a leadership position in developing cleaner and more efficient processes that industry in other countries will be anxious to buy from them. Steps to reduce our dependence on imported oil are so essential for other reasons, as evidenced by the Iraqi invasion of Kuwait in the summer of 1990, that the Federal government should take much more expansive actions than are currently in place to reduce that dependency and ease the nation's trade deficit. The fall 1990 budget agreement to raise the federal tax on gasoline by five cents is too small to have much of an impact on energy use, and the failure of Congress in 1990 to increase the corporate average fuel economy (CAFE) standards set back efforts to reduce oil consumption.

The federal government provided almost $2.86 billion in assistance for urban mass transportation systems in 1989. By 1991, that figure is expected to drop to $2.48 billion as the Bush administration has sought to reduce the federal role in financing transportation projects. The administration proposed to eliminate funds to help large cities operate mass transit systems and to only fund new projects for which the federal government has already signed contracts (U.S. Department of Transportation 1990).

Funds for the development of alternative fuels could be multiplied by several times in order to provide a major stimulus to developing alternative energy sources. If subsidy levels matched those given to nuclear power, for example, tremendous progress could be made in developing new sources of energy. Renewable fuels, including solar energy, biomass, geothermal, hydro, and a number of other kinds of fuels, promise to provide a virtually inexhaustible supply of energy if carefully managed. These fuels vary in their environmental impact, costs, energy efficiency, and availability, but in most cases they have clear advantages over fossil fuels. Hydroelectric power, currently producing about seven percent of the world's commercial energy supply, is growing most rapidly in the developing world, but there are major economic and environmental barriers to increasing hydroelectric power production. Solar thermal and photovoltaic energy technologies are widely used throughout the world. Wind power has been harnessed in some areas at costs that approach conventional sources of electric power. As the relatively high costs of producing energy with these technologies are reduced, their use will become more widespread. Other sources of energy that rely on new technologies, such as orbiting solar collectors and power plants that use the temperature differences between warm surface temperature and cooler, deeper ocean waters to produce energy, require more development before becoming commercially feasible (Global Tomorrow Coalition 1990, 197–98).

Seventh, both public and private spending on basic investments in the infrastructure of economic activity needs to increase. The level of savings in the U.S. plays a major role in determining the availability of resources for investment. Gross domestic saving in the U.S. (the sum of personal, business, and government savings) averaged 16.5 percent of the GNP during the 1960s and 70s, but fell to 13.2 percent in 1988. The decline in U.S. savings was offset by increased foreign investment in the U.S., a short run boost but a long run liability as future interest and dividend payments will flow overseas. Personal savings as a percentage of GNP fell from a 5.6 percent average in the 1970s to a low of 2.2 percent in 1987 and only increased slightly in subsequent years.

The low rate of savings in the U.S. is the result of several factors. Social Security may dampen incentives to save because of the resources it provides for retirement that is otherwise a major reason for personal savings. High interest rates and changes in the tax code in 1981 were expected to have a positive impact on savings that have not occurred. The increase in the late 1970s and 80s in the number of people in the peak buying and borrowing phase of their lives—the baby boom generation—resulted in lower savings rates. The appreciation of the value of homes has skewed calculations of savings, making rates appear lower than they otherwise might be. Capital gains are taxed more heavily in the U.S. than many other industrial nations. Interest on income is fully taxed in the U.S. but not in Japan, for example. Dividends are taxed at both the corporate and individual level (U.S. O.M.B. 1990, 45). The overall savings rate could be increased by a number of actions, ranging from budget deficit reductions to using the Social Security system's surplus to retire part of the national debt to Family Savings Accounts and enhanced IRAs that would feature tax-exempt interest. While these issues affect investment activity, they are part of a much broader political and economic context in which investment for competitiveness is only one of several competing concerns.

Conclusion

Debates over competitiveness often conclude with the concern that the primary problem is the tendency to evolve into assistance to declining industries that are no longer competitive. That is a problem that cannot be ignored, given the history of pork barrel politics in Congress and the continual temptation of members of Congress to circumvent the procedures they established or required others to do in earmarking projects for their own districts. However, the risk of not taking sufficient action to promote competitiveness is greater than the risk of perpetuating some pork barrelling. The far greater challenge is for Congress and the president to develop the political will to make investments in the long run—to impose short run costs, such as reducing the budget deficit and increasing investment, in order to achieve long-run benefits. The sweeping changes in the global economy, new national security challenges, and the growing importance of protecting natural resources and environmental quality provide the context in which competitiveness policy must be made. These factors simply overwhelm the preoccupation in the U.S. with the bailouts of declining industries and the ideological debate over "free markets."

Given the dynamic nature of the global economy, there is considerable risk in doing nothing, in failing to take initial steps to improve the base on which economic competitiveness rests. Congress and the executive branch need to engage in a continual, dynamic assessment of the strengths and weaknesses of alternatives for encouraging competitiveness. Office of Technology Assessment studies, business analyses, academic reports, and other sources of information on emerging technologies are available. The two institutions need to devise institutional arrangements that permit them to work together in fashioning policies that are flexible and able to be changed as conditions warrant adjustments. Policies need to be viewed as catalysts to changes in the private sector rather than direct federal involvement.

The Federal government has for decades maintained an "industrial policy" of assistance to industries to ensure a strong national defense. While some poor investments have been made, there have been some significant successes. The challenge now is to make more explicit our commitment to providing the preconditions for effective commercial and industrial activity, and not simply assume that they are produced as a by-product of ensuring national security. Investments that contribute to long-run economic viability and that will provide a strong base for scientific progress and its commercial application ought to be among our highest public priorities. A willingness to take risks, coupled with a real commitment to learn from experience and make adjustments accordingly is critical. Poor policy choices will continue to be made, but that risk is less dangerous than the failure to ensure that government makes strategic contributions to the viability of the U.S. economy.

Bibliography

Brandon, Richard N. October 1990. "Expert Advice for Congress on Science and Technology Budget Decisions." Paper presented at the Annual Meeting of the Association for Public Policy and Management, San Franciso.

Berger, Stephen. 10 June 1990. "The Cookie Monster Makes Policy," *New York Times*.

Boskin, Michael. 5 February 1990. John Carey and Douglas Harbrecht, "The Future of Silicon Valley." *Business Week*: 60.

Brooks, Harvey and Lewis Branscomb. August-September 1989. "Rethinking the Military's Role in the Economy." *Technology Review*: 55–64.

Browning, Graeme, 6 July 1991. "Techies in Cahoot." *National Journal*: 1687–90.

Carey, John. 15 June 1990. "United States." *Business Week*: 48.

Chubb, John E. and Terry M. Moe. 1990. *Politics, Markets, and America's Schools*. Washington D.C.: Brookings Institution.

Congressional Quarterly. 1988. *Federal Regulatory Directory*. 5th ed. Washington D.C.: Congressional Quarterly.

Department of Commerce. *Competitive Status*: 10–14.

Dertouzos, Michael L., Richard K. Lester, and Robert M. Solow. 1989. *Made in America: Regaining the Productive Edge*. Cambridge, Mass.: MIT Press.

Executive Office of the President. Office of Management and Budget. 1990. *Budget of the United States Government, Fiscal Year 1991*. Washington D.C.: GPO.

Fallows, James. 1982. *National Defense*. New York: Vintage.

Gephardt, Richard. April 1990. *The Competitive Status of the U.S. Electronics Sector*. Washington D.C.: GPO.

Global Tomorrow Coalition. *Global Ecology Handbook*.

———. *Global Ecology Handbook*.

Huber, Peter. 1988. *Liability: The Legal Revolution and Its Consequences*. New York: Basic Books.

Johnston, William J. ed. 1985. *Education on Trial: Strategies for the Future*. San Francisco: Institute for Contemporary Studies.

Lamm, Richard D., Richard A. Caldwell, and Ira H. Mehlman. 1989. "Hard Choices." Denver: The Center for Public Policy and Contemporary Issues.

Lawrence, Robert Z. 1990. "Innovation and Trade: Meeting the Foreign Challenge." In *Setting National Priorities: Policy for the Nineties*, edited by Henry J. Aaron, 146–52. Washington D.C.: Brookings Institution.

Magaziner, Ira C. and Robert B. Reich. 1982. *Minding America's Business: The Decline and Rise of the American Economy*. New York: Vintage.

Mitchell, Robert Cameron. 1990. "Public Opinion and the Green Lobby: Poised for the 1990s." In *Environmental Policy in the 1990s*, edited by Norman J. Vig and Michael E Kraft, 81–102. Washington D.C.: Congressional Quarterly Press.

Nanto, Dick K. and Glean J. McLoughlin. September 1990. "Industrial Associations and High-Technology Policy Making." *LRS Review*: 32–33.

National Commission on Excellence in Education. U.S. Department of Education. 1983. *A Nation at Risk.*

Osborne, David. Summer 1990. "Refining State Technology Programs." *Issues in Science and Technology*: 55–61.

Porter, Michael. 1990a. *The Competitive Advantage of Nations.* New York: Free Press.

———. March-April 1990b. "The Competitive Advantage of Nations." *Harvard Business Review*: 81–3.

Rubin, Alissa. 2 March 1991. "Proponents of Technology R&D Emphasize Competitiveness." *Congressional Quarterly Weekly Report*: 534–36.

Schlefer, Jonathan. August-September 1989. "Making Sense of the Productivity Debate: A Reflection on the MIT Report." *Technology Review*: 31–33.

Stubbing, Richard A. 1986. *The Defense Game.* New York: Harper and Row.

Sullivan, Allanna. 17 August 1990. "It Wouldn't Be Easy, But U.S. Could Ease Reliance on Arab Oil." *Wall Street Journal*: 1.

U.S. Congress. Office of Technology Assessment. June 1988. *Paying the Bill: Manufacturing and America's Trade Deficit.* Washington D.C.: GPO.

———. February 1990. *Making Things Better: Competing in Manufacturing.* Washington D.C.: GPO.

U.S. Department of Commerce. April 1990. *The Competitive Status of the U.S. Electronics Sector.* Washington D.C.: GPO.

———. 15 June 1990. *Business Week*: 12–30.

U.S. Department of Energy. 1990. *Interim Report: National Energy Strategy.* Washington D.C.: Department of Energy.

U.S. Department of Transportation. 1990. *Moving America: New Directions, New Opportunities.* Washington D.C.: U.S. DOT.

U.S. Office of Management and Budget. 1985–1988. *Regulatory Program of the United States Government.* Washington D.C.: GPO.

———. 1990, 1991. *Budget of the United States Government, Fiscal Years 1990, 1991.* Washington D.C.: GPO.

Waldholz, Michael. 24 August 1990. "Colon Cancer Growth is Halted in Tests by Replacing 'Tumor Suppressor' Genes." *Wall Street Journal*: 1. World Resources Institute. 1988. *World Resources 1988–89*: 114. New York: Basic Books.

Young, John A. 15 July 1988. "Technology and Competitiveness: A Key to the Economic Future of the United States." *Science*: 314.

PART THREE

Assessing Policy Analysis:
The Challenges
of Technology Policy

11

Parliamentary Technology Assessment in Europe: A Comparative Perspective

Norman J. Vig

The New Europe of the 1990s has a decidedly hi-tech look to it. With the achievement of a single economic market in 1992, the twelve nations of the European Community (EC) will begin a new era of competition and innovation. Among the goals of this historic project is the launching of new strategic industries based on "core technologies" such as microelectronics, computers, robotics, telecommunications, biotechnology, and advanced materials. The EC already supports a formidable array of collaborative research and technological development (R&TD) programs to prepare for the global technological competition of the future (Sharp and Shearman 1987; European Communities 1988; de Woot 1990). But each of the industrial nations spends far more on its own science and technology policies in what has been called the "technology race among OECD countries" (Roobeek 1990).

This emphasis on the economic importance of science and technology is not new in Europe, but it has taken some three decades to move to the top of the agenda. Two of the original European community organizations—the Coal and Steel Community (1950) and Euratom (1958)—were created to modernize and develop basic industries. The Council of Europe (1949) and Organization for Economic Co-operation and Development (1960) took an early interest in the development of national science policies, and during the 1960s most member governments created new ministries or other bodies to develop such policies (OECD 1963, 1971). For the first time, political conflicts arose in a number of countries over both the level of government R&D expenditure and criteria for allocating scientific resources (Vig 1968; Long and Wright 1975). In France, Jean Jacques Servan-Schreiber warned darkly of U.S. technological hegemony unless Europe pooled her resources to compete (Servan-Schreiber 1968). It was in this period that the first large European collaborative projects were undertaken in such fields as aviation, space, and advanced nuclear reactors and that the concept of an integrated European "technology community" emerged (Layton 1969; Sharp and Shearman 1987; Sharp 1989).

The 1970s raised new challenges. While European cooperation floundered amidst economic crises and intergovernmental disputes over EC policies, the societal consequences of big science and technology came to be questioned by a wide range of new political groups. The New Left student revolt, the anti-war and anti-nuclear movements, and environmental groups all attacked the destructive impacts of technological development. Citizens demanded rights to participate in making technological decisions that affected their lives, whether in the form of nuclear power plants, airports, or toxic wastes (OECD 1979).

One consequence was recognition of the need for technology assessment (TA), the subject of this chapter. In 1972, the year in which the Technology Assessment Act was passed in the United States, the OECD undertook its first major study of TA concepts and their potential application in Europe (Hetman 1973). Debates also began within member states over how to institutionalize TA functions within their governmental systems, but, for reasons to be explained shortly, little progress was made by the end of the decade.

It was not until the 1980s that both technology policy and technology assessment achieved definitive governmental support in European countries. A flood of Japanese and Asian imports, the second oil shock, the deep recession of 1980–83, mounting ecological problems, and widespread protests and demonstrations over the emplacement of a new generation of nuclear missiles in Europe, all combined to force governments to consider both the positive and negative sides of technology. On the positive side, a new generation of "core" industrial technologies came to be seen as the answer to the stagnation and unemployment of the previous decade. "Industrial revival through technology," as the title of a recent OECD publication put it, became the order of the day (OECD 1988a). At the same time, it was recognized that technological development is a social process requiring both structural economic reforms to encourage innovation and social and political reforms to ensure that such development has adequate human resources and public support (OECD 1983, 1988b). Aspects of technological change such as labor market alterations and new education and training requirements must be treated as an integral part of technology policy: "a strategy for technology is more than a policy for R&D" (OECD 1988b, 23).

"Constructive technology assessment" is therefore needed to evaluate and guide the development of new technologies and to ensure their public acceptance. The following recommendation of a recent OECD report perhaps best summarizes the European perspective on technology assessment:

> Our starting point was that technological change is a social process. From that follows the need for a broad-based consensus about the impact of new technologies on the social fabric, both at the national and international level. We therefore recommend the further development of various forms of technology assessment which should be a continuing process and in which elected legislatures should equip themselves to play an active and informed though not an exclusive role. The basic aim should be to provide information to those concerned, to promote and participate in a constructive public debate in a wide circle of institutions, thereby strengthening the democratic process through increased public understanding of, and involvement in, the process of change (OECD 1988b, 25).

The Role of Parliaments in Technology Policy

Legislative involvement in science and technology policy goes back more than a century. The U.S. Congress chartered the National Academy of Sciences in 1863 and commissioned a vigorous study of national technical needs in the 1880s (Guston 1990). In Europe government patronage of science long predated this, but parliamentary bodies were slow to gain authority and expertise over such matters. One of the oldest parliamentary bodies dedicated to informing members of scientific and technological advances is the Parliamentary and Scientific Committee in the British Parliament, a loose grouping of MPs, peers, and representatives of scientific organizations founded in 1939 (Walkland 1964). However, in Britain and Europe generally, it was not until the 1960s that parliaments began to take an active and sustained interest in science policy (Council of Europe and OECD 1965; Vig and Walkland 1966). In some cases this was institutionalized in the form of new parliamentary committees for science, research and/or technology which paralleled new ministries established at this time (e.g., the Select Committee on Science and Technology created by the British Parliament in 1966 and the Committee on Research and Technology of the West German Bundestag set up in 1972).

These committees, while having some powers to take testimony and collect evidence on government R&D programs and to issue reports focusing on particular needs or shortcomings, generally lacked authority to influence policy. The reason is not hard to find: in parliamentary systems the prime minister and cabinet govern, not the legislature. They determine all important policy, propose legislation, and control the budget. Parliament itself is organized along party lines to ensure the control of the majority or governing party over all proceedings. In contrast to our system, European parliaments have few independent powers to challenge the executive (short of sacking the government) and, instead, concentrate on debating and implementing its policies. Even this can be difficult since the government is usually loath to share information that might benefit the opposition. Thus it is very difficult for parliamentary committees to conduct their own investigations or to establish a claim to their own sources of information.

For these fundamental reasons, it has been extremely difficult to develop the function of technology assessment in parliamentary settings. In particular, it has been impossible to transfer the institutional structure of the U.S. Office of Technology Assessment to European countries despite many efforts to do so. It must be remembered that OTA was established in large part to provide Congress with independent and impartial technical information that would allow it to question executive branch proposals. Although there were other purposes as well, such as to provide long-range warnings of potential adverse consequences of new or developing technologies, OTA was designed to serve and enhance the institutional prerogatives of Congress. Its structure, especially that of the bipartisan Technology Assessment Board which screens committee requests for project studies and buffers the staff from political interference, reflects the unique autonomy and individualism of congressional decisionmaking (Gibbons and Gwyn 1988). No European parliament has such independence and freedom from partisanship, and thus TA initially developed as an executive rather than a parliamentary function in Europe.

It is not surprising therefore that a decade after OTA was founded one of its leading experts could conclude that:

> The United States is still the only nation that has a technology assessment support organization to serve the national legislature. There have been a number of attempts to establish such a body in other nations, including Sweden, the Netherlands, Germany, the United Kingdom, and France. These efforts have all failed. . . . There is a growing consensus that technology assessment functions cannot be effectively implemented as a legislative support in many countries (Coates and Fabian 1982, 343–44).

But, to paraphrase the immortal words of Mark Twain, it now appears that reports of technology assessment's death were exaggerated. In the past half dozen years, all of the aforementioned countries save Sweden have established new parliamentary TA mechanisms, as have Denmark and the European Parliament itself. None are comparable to OTA, but they are beginning to provide a basis for greater parliamentary involvement in technology decision making.

The Relaunching of Technology Assessment

Technology assessment was debated in Europe for more than a decade before institutional solutions began to emerge. The difficulties of reaching a conclusion were perhaps most manifest in West Germany, where it took 17 years of discussion, at least 15 separate proposals, and two special study commissions before an institutional mechanism was finally agreed upon in 1990 (see below). In Britain the debate lasted almost as long before the Parliamentary and Scientific Committee formally proposed a technology assessment office to Mrs. Thatcher in 1986. However, the Prime Minister refused to give it financial support. As a result, the Parliamentary Office of Science and Technology (POST) had to be set up as a charitable trust outside Parliament in 1988. In France an *Office d'Evaluation des Choix Scientifique et Technologiques* was created in 1984; Denmark established a parliamentary Technology Board in 1986; and the Netherlands Organization for Technology Assessment—whose research program must be approved by the Dutch parliament—was begun the same year. In other countries such as Austria and Sweden TA has been institutionalized in units that remain unattached to parliament, while in still others such as Italy and Spain, plans for parliamentary TA are in the early stages.

In 1987 the European Parliament in Strasbourg launched the Scientific and Technological Options Assessment (STOA) program to inform MEPs on European-wide technology issues and related European Commission proposals (McBrierty 1988). The STOA office also maintains liaison with the parliamentary TA bodies and consulting experts in several countries.

One of the reasons the parliamentary TA proposals failed initially is that they followed the American model too closely, often reflecting visits by parliamentary delegations to Washington (Coates and Fabian 1982). Another is that they were usually initiated by the opposition parties, almost automatically guaranteeing rejection. In Germany, for example, the CDU/CSU minority pushed for TA when the Social Democrats were in power, but from 1982 it was the SPD and Greens who advocated TA against the wishes of the CDU/CSU government. In France conservative governments blocked action until 1981 when election of Mitterrand's Socialist Party opened the way. In Britain the Conservatives have lately been the stronger advocates, but they have not been able to convince their own government to support it. These and other national experiences are discussed further below.

In addition to overcoming constitutional and partisan hurdles, the *concept* of technology assessment had to evolve beyond its original formulations to find intellectual acceptance in Europe. As originally conceived, TA was mostly regarded as a negative tool for controlling runaway technology. In the intellectual climate of the late 1960s and early 1970s in America, this made sense: the unanticipated consequences of technologies seemed at the root of most social and environmental problems. The emphasis was thus on "early warning" against insidious incremental effects of new and developing technologies. Technology assessment, like environmental impact assessment, was supposed to help us to avoid such disasters in the future.

By the 1980s the intellectual climate in both the U.S. and Europe had changed considerably. Despite public fears about particular technologies such as nuclear power and nuclear weapons, technology in general was more apt to be viewed as part of the solution than as the source of the problem. This was partly due to rapid technological breakthroughs in computing, telecommunications, biotechnology and other fields which appeared to offer real prospects for economic growth, and partly the result of a growing perception that technology is a human creation whose direction and consequences can be influenced. But experience with technology assessment also suggested that the original concept of single, comprehensive, objective analyses of individual technologies was naive, and that a more realistic approach was therefore needed (Smits and Leyten 1988).

European students of TA such as Renate Mayntz had observed for some time that decisionmaking for technology "resembles a diffused process of negotiation in which discrete decisions can hardly be distinguished an din which political rationality is more important than scientific rationality" (quoted in Smits and Leyten 1988, 28). If this is so, decision makers are unlikely to be influenced by purely technical analysis, especially one-time "reactive" reports on specific technologies or projects. TA should rather be an ongoing, "active" process involving broad social dialogue on the values embedded in alternative kinds of technological development (i.e., at the strategic level of choice rather than only the individual or project level). On the one hand, this implies an effort to design beneficial technologies—what the Dutch call "constructive technology assessment." On the other hand it suggests a need for new institutional processes to promote broader public discussion and citizen participation in technology assessment. This also implies that TA should be a pluralistic process, with parliament facilitating and aggregating research and discussion carried out in multiple institutional settings and networks (universities, corporations, local governments, research institutes, etc.) rather than attempting to produce definitive reports of its own.

This broadened approach to technology assessment is clearly more suited to the role of parliaments than the more policy specific and technically oriented OTA model. It recognizes the primacy of political debate and leaves room for the interjection of differing cultural values in to the TA process. Arguably, it also reflects a more sophisticated and realistic understanding of the social and political foundations of technological choice than our more analytical approach to evaluation. The results are likely to bring out the inherent uncertainties and complexities of technological change and may even increase the difficulties of making social choices. Nevertheless, the European approach may produce more socially beneficial outcomes in the long run.

So far, though, conceptualization of TA has run well ahead of practice in Europe (the opposite might be said of the U.S.). The following section provides examples of recent efforts to operationaize TA in several countries.

Federal Republic of Germany

Since the TA debate has gone on longest in Germany and has been documented most fully, we can begin with the FRG as an example of the difficulties of institutional transfer. The first proposal to establish an OTA-type body (an "Agency for the Assessment of Technological Developments") in the Bundestag was made by the CDU/CSU Fraktion (parliamentary group) in April 1973, following indications from the newly elected SPD/FDP government of Willy Brandt that it was open to the concept of social impact assessment of major projects. Discussions had been held in the Bundestag's *Committee on Research and Technology*—itself just established to shadow the new *Ministry of Research and Technology* created in 1972—where it became evident that the "ministerial bureaucracy was not very enthusiastic about the existence of a group of experts outside its responsibility" (Jaeger and Scholz 1990, 476–77). The government subsequently rejected the CDU/CSU proposal on grounds that the new body would be "technocratic" and lacking in democratic legitimacy; it was parliament's job to debate the broad goals and purposes of policy, but the government's to provide the specific resources (Jaeger and Scholz 1990, 477). In 1975 the government again rejected a CDU/CSU proposal, this time for a "Commission on Technology Assessment" to be comprised of six parliamentarians (equally divided between government and opposition parties) and six nonvoting experts. This was defeated on grounds of violating parliamentary rules on the composition of committees. However, the government also rejected a CDU/CSU proposal in 1977 to set up a small technology forecasting and evaluation staff under a coordinating council reflecting proportional party strength, and an all-party proposal from the Committee on Research and Technology to create a small assessment unit in the office of the Bundestag president in 1978–79 (in this instance the parliamentary Budget Committee refused to allocate funds). The CDU group then proposed a steering committee under the guidance of both the Research and Technology Committee and the Budget Committee, but this, too, failed in 1981, and no agreement was reached before the Schmidt government lost its majority the following year.

This failure during the first decade of debate reflected the hard reality that, despite recognition in the abstract among all parties that assessment of the social impacts of technology is necessary, and that the Bundestag's role should be strengthened, no consensus emerged on either the concept of technology assessment or the principle that parliament should develop sufficient expertise to exercise control over government programs (Paschen 1987). This in turn reflected the weakness of parliamentary committees in the Bundestag, in this case the Committee on Research and Technology, which lacked any real legislative or budgetary powers. Thus even if the committee members could agree on an advisory support mechanism, they had no way of implementing it; and it was not obvious in any case that the committee should be the principal "client" or user of whatever TA might be done. By this time the government itself (primarily through its ministries for Research and Technology and Environment and Nuclear Safety) was already contracting out for technology assessments and saw no benefit in channeling additional TA funds through the Bundestag. Indeed it was easy to warn, even if facetiously, against creating a new "bureaucracy" for TA in parliament.

To compound the growing frustration and cynicism among the small group of parliamentarians, academics, and experts that consistently supported TA, the parties reversed their positions after 1982 so that the SPD now supported the ideas it had opposed in office while Helmut Kohl's new CDU/FDP government rejected its own former proposals. The impasse only began to be broken in early 1985 when the Bundestag finally accepted a proposal of the Research and Technology Committee to set up a special study commission (Enquete Commission) on "Assessment and Evaluation of Impacts of Technology; Shaping the Framework for Technological Development."

Study commissions of this kind are a unique feature of the German Bundestag. Their purpose is described as follows:

> Study commissions are a specific form of parliamentary policy deliberations. As a rule, half of the members of these commissions are Members of Parliament, while the other half are experts. The German Bundestag establishes such commissions in order to study complex and long-term issues and

to prepare decisions to be taken by the German Bundestag in the commissions' fields of work (German Bundestag 1989, 5).

These commissions, which expire at the end of each parliamentary term, have recently been utilized to investigate a number of major scientific and environmental issues, such as the future of the nuclear power industry, biotechnology development, and global climate change. Commission reports include recommendations for action and in some cases have had a major impact on public debate and on governmental policy. For example, the recent study on *Protecting the Earth's Atmosphere: an International Challenge* was adopted unanimously by the Bundestag and is being implemented by the federal government and used as a basis for international negotiations (German Bundestag 1989, 6).

The first Enquete Commission on technology assessment had much less impact, however. It recommended establishment of a permanent body much like itself—a commission comprised of parliamentarians and experts—to set the agenda and guide TA projects, which would be carried out by a scientific unit to be integrated into parliament. This unit was to subcontract the TA projects and monitor them, though it would also carry out some analysis itself. Advisory panels comprised of MP's and representatives of the public were to monitor the individual TA processes (Paschen et al. 1990). Despite a dozen years of debate, this proposal still bore striking resemblance to the original OTA design, the main differences being that outside scientists would have been incorporated into the steering apparatus while members of parliament would join the public in monitoring the projects.

Not surprisingly, the first Enquete Commission report was criticized both inside and outside the Bundestag. It aroused little interest among members, including the Committee on Research and Technology. Business groups meanwhile "expressed reservations about the proposal's consistency with a market economy and criticized the organizational form of the political steering group" because it included non-parliamentarians (Paschen et al. 1990, 5). No action resulted before the end of the tenth Bundestag in 1987.

A second Enquete Commission on technology assessment and evaluation was set up in the eleventh Bundestag session in late 1987 to conduct a number of TA studies and to reconsider the question of permanent arrangements. This group debated several organizational models put forward by the different party groups and finally produced a successful compromise. This plan makes the parliamentary research and technology committee (now renamed the Committee for Research, Technology and Technology Assessment) the political steering committee and client for a new TA institution that will operate both inside and outside parliament.

The committee will select topics, authorize projects, control finances, and release reports to the public (other committees can also propose topics to it). Much of the actual research and analysis will be done under contract by an independent research center, the Institute for Applied Systems Analysis (AFAS) at the Nuclear Research Center (Kernforschungszentrum) at Karlsruhe. AFAS will also set up and manage a small parliamentary office, the *Technikfolgenabschätzungsbüro Deutscher Bundestag* (TAB), which will function as a "discussion partner" and "translator" of TA findings to Bundestag members. Finally, the AFAS scientists are to serve as "mediators and coordinators" between scientific organizations and community groups in order to facilitate public participation in the TA process (Melder 1990; Jaeger 1990).

The new arrangement thus maintains the "primacy of politics" in that control of the TA process is vested in a parliamentary committee with majority representation from the governing party. Unlike the OTA in Washington, therefore, the ruling party will dominate the technology assessment "board." However, there are guarantees that a certain proportion of the committee budget for TA will be reserved for studies suggested by the minority parties (Schevitz 1990). Moreover, the external research unit will maintain strict scientific independence.

The director of AFAS and TAB, Dr. Herbert Paschen, was a member of the Enquete Commission and is perhaps Germany's leading expert on technology assessment. He takes a broad and flexible view of TA, arguing that it requires "every imaginable method" from virtually all disciplines, and that assessments should be conceived as ongoing, iterative "processes of intense interaction between the analysis team, the employer and other users or interested parties" (Paschen 1990b). AFAS has considerable past experience in TA research for the Ministry of Research and Technology and is widely respected for scientific objectivity. In its new parliamentary role it expects to add about ten scientists to a staff of 30 and to operate on a budget of DM 5–6 million per year. Its initial agenda includes two long-term studies—waste disposal and groundwater contamination—and several shorter projects on such topics as genetic screening, the space plane, and hydrogen fuel (Paschen 1990a).

AFAS's initial contract is for three years. If successful, it may well be that TAB is formally separated from AFAS with some of the AFAS staff moving permanently or being delegated to TAB for an extended period, thereby integrating TA as fully into the Bundestag as OTA is in Congress. This will depend, of course, on the quality of the assessments and their perceived usefulness to parliament and the public. More broadly, it will hinge on the ability of the new arrangement to reconcile the heretofore conflicting demands of party politics, government control, scientific research, and public concern over rapid technological change and environmental deterioration, all of which have been rendered even more pressing by the reunification of Germany.

The United Kingdom

The British Parliamentary Office of Science and Technology (POST) is situated even more precariously than its German counterpart since it is not formally part of parliament at all. In 1986 a group of MP's and peers met with Margaret Thatcher to discuss establishment of a Westminister version of OTA, but the Prime Minister, while not opposed to the idea in principle, refused to grant public funds. This was consistent with her effort to reduce the size of government by eliminating unnecessary commissions, committees, and "quangos" (quasi-nongovernmental bodies). She proposed instead that the Parliamentary and Scientific Committee set up an unofficial information office with private funding; and a charitable trust fund was subsequently created with support from individual MP's, companies, professional associations, and foundations. In office space donated by the Society of Engineers near Westminster Abbey, POST began operations in April 1989 with a Director and part-time secretary. Since then its staff has grown to five.

Institutionalization of parliamentary TA faced somewhat different hurdles than in Germany. There is a long British tradition of appointing nonpolitical committees or royal commissions to study complex technical issues (Baker 1988; Coates and Fabian 1982). Within parliament, matters of science and technology were likewise regarded as nonpartisan matters, giving rise to the cross-bench Parliamentary and Scientific Committee and, in 1966, to a new Select Committee on Science and Technology. Select committees are explicitly nonpartisan and non-legislatively oriented; their job is to scrutinize public administration to improve service and efficiency (Vig and Walkland 1966). After the 1979 election, the system was reorganized by the Conservatives and a new set of select committees was set up in the House of Commons to shadow each of the major departments. Thus new Commons committees were created *inter alia* for Education and Science, Energy, and Environment. The House of Lords, however, chose to retain its own Select Committee on Science and Technology.

The 1980s was a decade of considerable turmoil for British science, especially university research, due to budgetary cutbacks by the Thatcher government. Funding for civilian R&D lagged even further behind defense, raising new questions about Britain's ability to compete in high technology markets (Williams 1988). The Lords committee on Science and Technology played a leading role in criticizing the government's misuse of R&D statistics to bolster its case. Meanwhile, the Commons select committees conducted a number of inquiries into controversial issues such as radioactive waste disposal (O'Riordan 1988). In this contentious climate, the government may have been especially wary of establishing any general capacity for technology assessment within parliament.

POST has been scrupulously nonpartisan in orientation. It defines its function as providing accurate, objective scientific information to MP's to improve their understanding of issues and raise the quality of parliamentary debates (Norton 1990). Its director, Dr. Michael Norton, is a chemist who spend several years in the U.S. and became thoroughly familiar with OTA's organization and procedures. Perhaps more than in any other country, he has tried to adapt OTA's methods and philosophy to parliamentary conditions. Given resource limitations, however, this has required a drastically scaled-down version.

Like OTA, POST is supervised by a bipartisan board drawn from both houses, but the board also contains five distinguished nonparliamentary scientists and engineers who provide expert advice. Although both parties are represented—Sir Ian Lloyd, a Tory MP is chair, and Lord Kennet, a Labour peer, is vice-chair—the Conservatives have been most active thus far. The board meets quarterly with the director, approves research topics, and must approve completed documents before release.

The greatest difference between OTA and POST is that POST has had to limit its work to relatively small, short-term studies of immediate use. Aside from lack of staff, MP's are unlikely to be interested in large reports which take a year or two to produce and have little relevance to the parliamentary agenda (over which they have no control). Hence POST has concentrated on producing short "Briefing Notes" (about one per month) that cover subjects of current political interest such as *in vitro* fertilization and embryo research, drinking water quality, computer misuse, and release of genetically modified organisms into the environment (2). These Notes (2–4 pages each) provide basic objective information that has helped to focus parliamentary debate. MP's are also invited to seek further details from POST.

These briefings, while timely and useful, are not really technology assessments. However, POST launched four larger projects in 1990 that can be considered TAs: one on the use of new technologies in education (carried out by the Open University under POST direction); one on the interrelationships between defence R&D and civil science and technology; one on the environmental costs of fossil fuel energy use; and one on research in the National Health Service. The second and third of these studies are being conducted by outside scientists who are receiving most of their support from other sources. Most of the TA work within POST is done by three scientists on fellowships or secondment from charitable foundations and scientific societies. This arrangement is not sustainable for more than three or four years, by which time POST hopes to convince parliament and the government that it deserves permanent public funding.

Other National Experiences

Other countries have attempted to find a niche for parliamentary TA within their particular institutional contexts. One of the dilemmas suggested by the discussion thus far is that if the TA function is incorporated too fully into parliament it runs the risk of politicization and consequently resistance from the government (as has been the case until recently in Germany). On the other hand, if TA is too far removed from parliamentary decisionmaking it will lack any client for its services and simply remain irrelevant. A second issue is how effectively the TA office is in networking with expert groups outside parliament and thus providing a bridge to the scientific community. This is partly a function of a third factor—whether the parliamentary office has enough staff and resources to attain the "critical mass" necessary to conduct or organize major studies. And finally there is the question of whether technology assessment is sufficiently open to the concerns of citizens and social groups to gain their participation and support. So far none of the European TA experiments have resolved these issues.

The French have perhaps gone the farthest in attempting to incorporate the TA function into parliamentary processes. The *Office Parlementaire d'Evaluation des Choix Scientifiques et Technologiques* (OPECST) was established in 1984 after several years of debate between the National Assembly and Senate. The Socialist government of Francois Mitterrand elected in 1981 launched a vigorous new technology policy to reinvigorate French industry, but also favored strengthening the role of parliament (Papon 1988). An explicit purpose of OPECST was to break the monopoly of scientific information enjoyed by the executive offices and the large technological institutes, and to furnish parliament with sufficient information to evaluate and formulate technology policy itself. OPECST is formally comprised of a parliamentary delegation or committee of eight senators and eight deputies, who are in turn advised by a council of fifteen eminent scientists. The delegation selects topics, gathers evidence, and commissions studies (it conducts no research itself), but also drafts the final TA reports with the assistance of a small administrative staff of five. Proposals for TA studies may originate only from committees of the two houses.

The French model thus maximizes parliamentary independence by tying technology assessment closely to its committee structure and the prerogatives of both houses. However, this appears to be as much a source of weakness as of strength in the French system. The differing composition and traditions of the two houses have made cooperation difficult, while shifting party majorities since 1986 have made it difficult to avoid politicization. Although OPECST has carried out studies on such important topics as acid rain, the effects of the Chernobyl accident, the impact of AIDS, high-definition television, and the Antarctic Mining Convention, at least one French expert has concluded that "little attention is paid to its reports at present, and its future seems uncertain" (McBrierty 1988, 9).

In the Netherlands, on the other hand, there is greater emphasis on achieving consensus among government, parliament, and scientific advisory bodies. Although proposals for a "Dutch OTA" were first introduced in parliament in 1975, it was concerns over the social impacts of microelectronics development, genetic engineering, and nuclear energy that sparked a debate in the late seventies and early eighties which prompted the first Lubbers government to endorse creation of a TA body in 1982. This was initially to be located in a bureau of the Ministry of Education and Science, but after debate of the government's plans the Lower House of Parliament demanded "the creation of a greater distance between TA organization and executive power without, nevertheless, demanding a significant say for itself in the organization of TA" (Paschen et al. 1990b, 2). The result was a complicated compromise that established a limited parliamentary role in the Netherlands Organization for Technological Assessment (NOTA), created in June 1986.

NOTA's organization and functions have been succinctly outlined as follows:

NOTA consists of a steering committee of nine people and a bureau. The chairperson and four members are nominated by the Royal Dutch Academy of Sciences (KNAW). Four members are nominated by the WRR (Scientific Council for Government Policy). The steering committee is assisted by an executive bureau of four to six people (primarily those people who were involved earlier at the Ministry . . .).

NOTA works as follows: Every year, the bureau prepares a program of research and associated activities (workshops, conferences, discussions, etc.) for the following year; the steering committee assumes responsibility for this program and presents it to the Minister of Education and Science who transmits it, together with his comments, to the Lower House. The Lower House discusses the proposal and sends it back to the minister together with its commentary. He then settles, taking those comments into account, on the definitive program and sends it to NOTA to be carried out (Paschen et al. 1990b, 3).

NOTA guides the whole process from choosing topics to the writing of final reports to parliament, but two-thirds of the research work is contracted out to university groups and research institutes (van Eijndhoven, 1991).

NOTA's greatest contribution, perhaps, has been in propagating the idea of Constructive Technology Assessment (CTA), which the Dutch characterize as an active, positive form of shaping technological development as opposed to the original "early warning" approach (Schot 1992). Under this theme, NOTA has sponsored research and workshops on new technologies such as artificial intelligence, in-vitro fertilization, biotechnology and agriculture, and ISDN (Integrated Services Digital Networks). It also organized a large European TA conference in Amsterdam (Ministry of Education and Science 1987) and has been highly active in networking with sister organizations in other EC countries.

Despite these successes, NOTA has been criticized in parliament for initially spreading itself too thinly across a large number of small projects and for the timing and usefulness of its reports. Its relationship to parliament and outside social organizations is also ambiguous, since it continues to work closely with the Education and Science ministry. Parliament is not the only TA client as in the other cases we have discussed. Mechanisms for citizen participation have also languished after initial promises of broad social access. However, relations with business organizations have improved, and the Ministry of Economic Affairs has begun to show greater interest in CTA as a guide to industrial policy.

Indeed, with professional staffs of only about five persons each, it is questionable whether the TA offices of Britain, France and the Netherlands can conduct many assessments of the quality achieved by OTA, or significantly influence parliamentary deliberation. Considerably more funding is needed in all countries to build up pluralistic TA research capabilities and networks for broadened societal participation. Most importantly, TA practitioners have to cultivate stronger political ties to influential members and committees of their respective parliamentary bodies if they are to follow in OTA's footsteps.

Conclusion

It should be apparent from this brief review that while OTA is too highly adapted to its particular ecological niche in the U.S. Congress to survive direct institutional transfer, the general theory and practice of technology assessment is beginning to evolve along parallel, if not identical, lines in many European countries. This is not surprising since TA is part of a larger family of new techniques for comprehensive impact analysis that are being universalized as tools for decisionmaking (Bartlett 1989). Although the fusion of legislative and executive powers in parliamentary systems weakens the case for independent TA structures, parliamentarians realize that they cannot perform even traditional functions of representation and debate without understanding the social consequences of (and opportunities brought about by) new technologies. Thus, while recognizing that TA should take pluralistic forms, there is a growing movement in Europe to locate at least part of the TA function in parliament.

The newly unified Federal Republic of Germany has some of the most daunting technological problems to face, both as the leading industrial nation of the European Community and as the model for rebuilding the backward and polluted societies of what was Eastern Europe. The Bundestag's fledgling TA apparatus is perhaps the most promising in Europe and might well provide the model for the future. Other nations of the EC, as well as their newly democratic cousins to the east, will need to strengthen their parliamentary capacities if technology assessment is to contribute to a more humane and liveable, as well as prosperous and democratic, new order.

Bibliography

Baker, Richard. 1988. "Assessing Complex Technical Issues: Public Inquiries or Commissions." *Political Quarterly* 59(2): 178–189.

Bartlett, Robert V. ed. 1989. *Policy Through Impact Assessment: Institutionalized Analysis as a Policy Strategy.* New York: Greenwood Press.

Coats, Vary T., and Thecla Fabian. 1982. "Technology Assessment in Europe and Japan." *Technological Forecasting and Social Change* 22: 343–61.

Council of Europe and Organization for Economic Cooperation and Development. 1965. *Science and Parliament.* Paris: OECD.

de Woot, Philippe. 1990. *High Technology Europe: Strategic Issues for Global Competitiveness.* Oxford: Basil Blackwell.

European Communities. 1988. *Research and Technological Development Policy.* 3d ed. Luxembourg: Office for Official Publications of the European Communities.

German Bundestag, ed. 1989. *Protecting the Earth's Atmosphere: An International Challenge.* Bonn: German Bundestag.

Gibbons, John, H., and Holly L. Gwin. 1988. "Technology and Governance: The Development of the Office of Technology Assessment." In *Technology and Politics,* edited by Michael E. Kraft and Norman J. Vig, 98–122. Durham, N.C.: Duke University Press.

Guston, David H. 1990. "Congress and the History of Science: The Allison Commission, 1884–1886." Paper presented at the annual meeting of the American Political Science Association, San Francisco, 31 August.

Hetman, Francois. 1973. *Society and the Assessment of Technology.* Paris: OECD.

Jaeger, Dirk, and Peter Scholz. 1990. "Science and Technology in the German Bundestag Examined Through the Committee on Research and Technology." In *The U.S. Congress and the German Bundestag,* edited by Thaysen, Davidson and Livingston.

Jaeger, Dirk. 1990. "Ein Langer Weg im 'Hohen Haus'." *Das Parlament* 31 August–7 September, II: 36–37.

Layton, Christopher. 1969. *European Advanced Technology: A Programme for Integration*. London: Allen and Unwin.

Long, T. Dixon, and Christopher Wright, eds. 1975. *Science Policies of Industrial Nations*. New York: Praeger.

McBrierty, Vincent J. 1988. "Technology Assessment for Parliaments at National and European Level." *Futures* 20(1): 3–18.

Melder, Heinz-Joachim. 1990. "Die Vereinbarung Über die künftige Beratung." *Das Parlament*. 31 August–7 September, IV: 36–37.

Ministry of Education and Science. 1987. *Technology Assessment: An Opportunity for Europe*. The Hague: Dutch Ministry of Education and Science.

Norton, Michael. 1990. Personal interview. London, 23 June.

O'Riordan, Timothy. 1988. "The Prodigal Technology: Nuclear Power and Political Controversy." *Political Science Quarterly* 59(2), 161–77.

OECD. 1963. *Science, Economic Growth and Government Policy*. Paris: OECD.

———. 1971. *Science, Growth and Society: A New Perspective*. Paris: OECD.

———. 1979. *Technology on Trial: Public Participation in Decision-Making Related to Science and Technology*. Paris: OECD.

———. 1983. *Assessing the Impacts of Technology on Society*. Paris: OECD.

———. 1988a. *Industrial Revival Through Technology*. Paris: OECD.

———. 1988b. *New Technologies in the 1990s: A Socio-Economic Strategy*. Paris: OECD.

Papon, Pierre. 1988. "Science and Technology Policy in France, 1981–1986," *Minerva* 26(4), 493–511.

Parliamentary Office of Science and Technology. 1990. *First Annual Report, 1989–1990*. London: POST.

Paschen, Herbert, Thomas Petermann, Jeffrey Schevitz, and Ruud Smits. 1990a. "The Process of Institutionalizing Technology Assessment within the German Parliament." Forthcoming in *Research Policy*.

———. 1990b. "Technology Assessment in the Netherlands with Special Emphasis on NOTA." Forthcoming in *Research Policy*.

Paschen, Herbert. 1987. "Problems in Institutionalizing Technology Assessment: the Experience of the Federal Republic of Germany." *ATAS Bulletin* 4 (October): 36–39.

———. 1990a. Personal interview. Karlsruhe, Germany, July 2.

———. 1990b. "Was überhaupt ist Technikfolgenabschätzung?" *Das Parlament* 31 August–7 September, V: 36–37.

Roobeek, Annimieke J.M. 1990. *Beyond the Technology Race: An Analysis of Technology Policy in Seven Industrial Countries*. Amsterdam: Elsevier Science Publishers.

Schevitz, Jeffrey. 1990. Personal communication.

Schot, Johan W. 1992. "Constructive Technology Assessment and Technical Dynamics: The Case of Clean Technologies." *Science, Technology & Human Values* 17(1): 36–56.

Servan-Schreiber, Jean Jacques. 1968. *The American Challenge*. New York: Atheneum.

Sharp, Margaret and Claire Shearman. 1987. *European Technological Collaboration*. London: Routledge & Kegan Paul.

Sharp, Margaret. 1989. "The Community and New Technologies." In *The European Community and the Challenge of the Future*, edited by Juliet Lodge. New York: St. Martin's Press.

Smits, Ruud, and Jos Leyten. 1988. "Key Issues in the Institutionalization of Technology Assessment: Development of Technology Assessment in Five European Countries and the USA." *Futures* 20(1): 19–36.

Thaysen, Uwe, Roger H. Davidson, and Robert Gerald Livingston, eds. 1990. *The U.S. Congress and the West German Bundestag*. Boulder, Colo.: Westview Press.

Tuininga, E.J. 1988. "Technology Assessment in Europe." *Futures* 20(1): 37–45.

van Eijndhoven, J.C.M. (NOTA Director). 1991. Personal interview. The Hague, 12 July.

Vig, Norman J. 1968. *Science and Technology in British Politics*. Oxford: Pergamon Press.

Vig, Norman J., and S.A. Walkland. 1966. "Science Policy, Science Administration and Parliamentary Reform." *Parliamentary Affairs* 19(3): 281–94.

Walkland, Stuart A. 1964. "Science and Parliament: the Origins and Influence of the Parliamentary and Scientific Committee." *Parliamentary Affairs* 17:3–4.

Williams, Roger. 1988. "UK Science and Technology Policy, Controversy and Advice." *Political Quarterly* 59(2): 132–44.

12

Toward More Effective
Policy Making for Science
and Technology

Gary C. Bryner

Congressional policy making is a regular target of academic, journalistic, and general public criticism. Few assessments of Congress fail to conclude that its members are unable to develop coherent policies, look beyond short-run political pressures, and escape an obsession with electoral consequences. While individual members of Congress are usually quite successful in providing the kinds of service their constituents seem to want, Congress, like other U.S. policy making institutions, seems largely unable to satisfy most of the expectations held for it (Chubb and Peterson 1989).

From a constitutional perspective, policy making in the American political system was designed to be inefficient, to check the power of policy makers, and to permit action only when there is widespread support among different sets of constituencies and their representative institutions. What was once defended as a virtue, however, is now regularly derided as a vice. Deliberative democracy has been replaced with deadlocked democratic processes (Burns 1984). Institutional barriers to coherent, unified, and effective policy making are believed by many to prevent the U.S. from responding to the challenges confronting the nation. Of particular concern is the fear that divided government—Congress and the White House being held by leaders of different political parties whose primary goal is the political defeat of the opposition—makes good, effective policy making extremely difficult at best, and often impossible (Sundquist 1986). Other institutional and structural characteristics of American government impede attempts to develop coherent, long-run strategies to resolve and prevent problems (Marshall 1987). The distribution of the consequences of technological advances seem to pose particular challenges to democratic policy making. Many of the adverse environmental consequences of industrial activity will fall on future generations, while the benefits are largely confined to the current generation. It is not clear how the interests of future generations or subgroups of the population that have the fewest economic and political resources can be protected in a political system dominated by well-heeled interests. Policy choices are permeated by uncertainty. There are often long lag times between policy interventions and adverse consequences. Decisions must be made by political actors, but they must be based on the latest scientific research. There is, however, rarely consensus in the scientific community and policy makers must function within a context of uncertainty and disagreement among experts (Jasanoff 1990). The nature of scientific debate and research, the slowness in which the scientific process often proceeds, and the tentativeness of scientific findings clash with short-run orientation of politics and the demand for unambiguous solutions to pressing problems. Science and technology-related issues cross traditional policy boundaries and disciplines. Despite the prominence of these features of science and technology policies, however, they are not unique and may simply accentuate the problems confronting congressional actions in other policy areas.

Public policies are by nature a function of political factors; political motivations, calculations, and concerns usually overwhelm policy analysis. The success of policies in accurately defining the nature and causes of problems and in developing and implementing effective solutions is a function of careful policy analysis as well as political acumen and luck. Central to the policy making process is the ability of policy makers to assess the strengths and weaknesses of alternative policy actions and to choose those that are most consonant with policy makers' values and priorities. Policy analysis in general rests on the expectation that the technical assessment of competing policy options will be separated from the political calculations of the policy makers, that there will be an objective, non-political assessment of policy options before the inevitable political calculations shape the decisions eventually made. Careful policy analysis will precede the application of narrow political pressure and ensure that policies producing the greatest net gains in social welfare will be pursued.

Policy analysis is not without its critics, however. The reliance of most analytic techniques on measures of economic efficiency and utility may clash with other values such as distributive justice. Reliance on analytic techniques may give the illusion of precision, certainty, and objectivity, when in fact decisions must be made on much more subjective grounds. Policy analysis may enhance the role of experts at the expense of elected officials, thus reducing the accountability of policy making to democratic forces (Hawkesworth 1988; Sartori 1962).

This book began by asking a number of questions. Can Congress fulfill these expectations for making policies in the face of uncertainty, or must we look more and more to the Executive Branch? Can Congress meet the challenges of making policies in areas involving complex scientific processes and technologies? Do members of Congress have sufficient knowledge to make decisions concerning the regulation of industrial and technological processes and their outcomes? How well suited are congressional institutions and processes to make policies based on these kinds of issues? The chapters above provide some very useful answers to these questions, but much more remains to be learned.

Challenges in Assessing Science and Technology Policies

As the chapters above demonstrate, science and technology issues are among the most difficult and complex challenges confronting congressional and other policy makers. The range of issues is staggering, from global climate change to genetic engineering, from international economic competitiveness to mapping the human genome. They often center on highly technical issues that are the subject of raging scientific debates or at the forefront of human knowledge. They implicate facts and theories that are sometimes only understood by a small number of leading scientists, or may not even be understood well by anyone. Yet Congress, with few scientists as members or staff, must commit the nation to profoundly important political choices.

Virtually every chapter pointed to the existence of a variety of analyses of scientific and technological issues prepared by congressional support agencies and others and made available to members of Congress. The challenge confronting policy makers was clearly not a lack of information, but were much more complex. Among the challenges identified were the following.

Value Conflicts

Policy decisions concerning science and technology are fundamentally choices among competing values, choices about how to act in the face of uncertainty, how to allocate scarce resources, how to insure against risks to human health, how to balance economic growth with other public interests, how to promote individual freedom and collective concerns, and how to satisfy the needs of the current generation along with those of generations to come. These choices are fraught with risks of colossal policy blunders and incalculable consequences for humankind as well as personal political dangers for those making the choices. These choices are largely inescapable; inaction commits societies to certain directions, despite efforts of policy makers to avoid decision making. Even good policy analysis provides little escape from these dilemmas.

As Woodhouse argues, there is no formula "capable of showing conclusively how tradeoffs should be made even for one individual, much less for a society." Blank's assessment of policy analysis for developments in biomedicine concluded that analytic failures were primarily a function of political and social constraints that result from the "high personal stakes that are inherent in the life and death issues surrounding biomedical policy." Particularly difficult in this policy area, but present in many others, is the challenge of giving representation to the interests of future generations and foreseeing how contemporary policy decisions will eventually affect them.

Regan demonstrated the importance of values such as personal privacy in a policy area dealing with complex computer and telecommunication technologies. While much of the attention focused on individual privacy, critical changes in bureaucratic power were occurring that were not well appreciated, and the concept of individual rights was really insufficient to address the range of values involved as new technologies emerged.

The primacy of values is, of course, no surprise since we are describing a political process and not simply a disinterested analytic one. Policy analysis must be infused with an appreciation of the political and social context in which policy making takes place. Among the greatest challenges in providing policy analysis is how to help political actors understand their choices and focus information that illuminates those choices while at the same time, as Woodhouse put it, resisting the naive belief that "more information per se will lead to a better world."

Uncertainty

Uncertainty has always been a part of congressional policy making. Demands for action are often made before problems are well understood. The consequences of taking legislative actions are difficult to predict. But science and technology issues raise the problem of uncertainty by orders of magnitude. The consequences of decisions can affect the entire planet, present, and future generations. They may commit us to irreversible repercussions. They require or at least assume familiarity with complex processes that most policy makers do not understand and have had little direct experience with. Perhaps most importantly, they make learning from experience, trial and error, incremental policy making and adjustment, very difficult. Woodhouse asserts that the "most significant task confronting those who would engage in technology policy analysis is dealing with the high level of factual uncertainty that usually characterizes development and diffusion of new technologies, and that plagues efforts to assess their social and environmental problems."

Traditional means of coping with uncertainty are taxed by the potential seriousness and long-run consequences of adverse outcomes, the long lead time and lag time before some risks are realized, and the interaction of complex and multiple causes that are often not well understood. Investments to promote economic competitiveness, for example, will take years before benefits are realized or failures become apparent. But resources to conduct policy analysis are limited, as is the attention span of policy makers; so a major challenge facing analysts and the in audiences is how to decide what to study and what kind of analyses would be most helpful. The challenge of policy analysis includes reducing uncertainties, through more and better data gathering and processing; perhaps more important, however, is helping decision makers cope with uncertainty, learn from experience, and make adjustments accordingly.

Information

Congress and policy makers do not suffer from a shortage of information and analyses. Yet often lack the kind of information they would like: concise, timely, oriented toward specific policy options, helpful in forecasting future societal problems and needs, sensitive to political and economic realities, and produced by individuals and organizations in whom they have trust and confidence. Major analyses from the Office of Technology Assessment, the National Academy of Science, academic scholars, and others may take too long to prepare or be unavailable to help with pressing legislative needs. Analyses by the General Accounting Office and Congressional Research Service lack the policy recommendations from politically significant interests that members of Congress rely on in helping to make their overall political decisions about complicated policy questions (Carnegie Commission 1991).

Wagner contends, however, that an inadequate scientific understanding of environmental issues is a major reason for the failure of environmental laws to achieve their goals. The role information and analysis plays in the policy process is a function of "the novelty of the problem for Congress and the degree of entrenchment of interests concerned with any policy solutions to it." But, ultimately, the real barriers to action are political, not technical in nature: "The technical complexity or scientific uncertainty surrounding a particular issue or aspects of it may increase the difficulty of policy making, but political willingness and other motivations of the key participants will be more determinative of the policy itself."

However, even when information is available, policy makers usually lack "the time, energy, and resources to digest more than a fraction of the information potentially available to them," as Woodhouse points out. They often "do not know how to conceptualize the problems in their domain of responsibility, or how to think strategically about them." If policy analysts are to produce more usable knowledge they must "explicitly address the actual conditions" facing policy makers.

Structure

The structure of congressional committees and subcommittees impinges on Congress's ability to address science and technology policy choices. Issues often cut across established committee jurisdictions. Efforts are often duplicative, as committees and subcommittees jealously guard their institutional prerogatives and maintain their turf. The decentralization of congressional authority raises tremendous challenges for comprehensive and coordinated policy making that some science and technology-related issues seem to require. More broadly, as Keller warns, the nature of divided government poses tremendous challenges to effective policy making. Given the wide range of policy failures, it is not at all clear that the fragmented U.S. political system can compete in the global arena "without adopting superior forms of technological and industrial organization."

The Range of Tasks

The range of responsibilities and tasks confronting Congress is daunting, overwhelming. Members must make decisions about what kinds of projects to fund, from large-scale, mega-projects to small-scale, decentralized research. They must choose between allocating resources themselves, or leaving those decisions to peer review and other processes. They must balance the need for support of basic science with demands for industrial and commercial applications of newly developed technologies. They must decide to promote and nurture some kinds of efforts while restrict and regulate others. They must mandate direct government intervention in some in economic and social enterprises, and provide incentives for other They must make legislation and then constantly monitor and oversee its implementation. A related challenge for policy makers is that public policy making represents only part of the decision making that affects the issues of concern. Congress and other government institutions are limited in terms of what they can accomplish with the incentives and sanctions at their disposal. As the chapter on competitiveness concludes, decisions made independent of government officials by others will play a major, and often a determinative role. Part of the policy-making challenge is for governments to serve as catalysts, to encourage private sector actions that are in harmony with public purposes.

The Role of Political Leadership

Policy analysis is often criticized for its failure to satisfy expectations that are unrealistic. Good policy analysis clarifies the strengths and limitations of alternative courses of action and identifies likely consequences. As Kraft maintains, good analysis can "inform political decision making," but "it is not a substitute for the judgment of public officials." Policy analysis is not necessarily inconsistent with politics: "it is only through a process of full policy legitimation—or democratic consensus building—that important political and technical issues, including potential obstacles to implementation, are likely to be understood and addressed effectively."

The case of nuclear waste policy provides a good example of a policy area where some excellent policy analysis has been done. Studies by the Office of Technology Assessment, the General Accounting Office, and House and Senate Committees provided Congress with abundant documentation concerning many of the scientific, administrative, and political issues surrounding nuclear waste disposal. Many members of Congress made considerable use of the information. The problem, according to Kraft, was "not lack of analysis, but insufficient use of it at critical junctures." Even the best analytic studies can be ignored or selectively used to reinforce positions that are, in reality, taken for other reasons. Political leadership is a critical ingredient. As Kraft concluded, Congress' ability to learn from and adjust to the Department of Energy's "aggressive and error-prone implementation of the 1982 act was short-circuited by Sen. Johnston's leadership style." Another style of leadership might have resulted in a somewhat different outcome.

Budget Conflicts and Funding Priorities

One of the greatest challenges for policy makers, particularly members of Congress, is to decide what resources to allocate to promote the development of science and technology. Not only are the initial decisions to launch government-funded projects difficult, but they must be periodically revisited in determining whether they should be modified or even canceled. These projects often involve high initial costs, long development times, and lots of uncertainties about their prospects. Kay's essay clearly captures the essence of this task:

> Successful development of these technologies requires a commitment of billions of dollars over a period of several years, or even decades, a fact which must be weighed against other budget priorities. In addition, because of their longer development times and inherent scientific uncertainties,

projects cannot be realistically evaluated until they are already well under way.

Policy analysis can provide policy makers with a sense of which facts are in dispute, but in this policy arena as well as most others, can also degenerate into a competition between "my experts" and "your experts." Advisory bodies and others can only provide limited help, Kay observes, "when the basic facts themselves are in doubt." As a consequence, "decisions regarding these technologies must of necessity be speculative in nature." Uncertainty is exacerbated by the dynamic nature of the policy agenda, as new issues emerge and others fade, and by shifts in the "attitudes, tastes, and perceptions of the public at large" that require Congress to revisit its decisions.

Scientists and engineers often express frustration, impatience, and even rage at congressional diffidence and inability to commit to a long-range agenda. Congress's role as a representative body inevitably produces policy uncertainty and change. Members of Congress must constantly reappraise spending decisions in light of new concerns and recalculations of competing values. As the nuclear fusion case demonstrates, scientists and engineers have been disappointed with the evolution of policy. Government, however, as Kay reminds them,

> does not develop technology solely for technology's sake, but rather seeks to identify those innovations that potentially can serve the public interest and to develop them in a socially productive fashion. Once an emerging technology comes under Congressional jurisdiction, it ceases being simply a "technical" problem, and becomes, for better or worse, a political issue. As such, it is prone to the same sorts of unexpected changes and priority shifts that can affect any other item on the public agenda.

Political decisions may impede the successful development of complex technologies and fail to serve the needs of the scientific community. The successful development of technologies like controlled nuclear fusion is a "formidable task even under ideal conditions. Continuous changes in program direction or levels of funding may make it impossible." It is not at all clear that the fractious politics of congressional policy making can produce expensive, complex technologies in a cost-effective and timely manner, that members of Congress can satisfy political principles and scientific and technological imperatives, and that the scientific and engineering efforts can flourish in such an environment.

The Evolution of Policies

Plein and Webber's discussion of biotechnology emphasizes how congressional technology assessment and information processing changes over time as the policy process develops. Given the importance and visibility in society of these emerging concerns, policy makers enjoy a wide range of sources of information concerning the various dimensions of biotechnology. The relative newness of this policy area, as well as the absence of entrenched, long-standing interests represented by congressional champions, gives congressional support agencies, particularly the Office of Technology Assessment, the opportunity to shape the basic contours of policy: "By developing policy knowledge, congressional support agencies become political actors in the policy process." Technology assessment done by OTA has already had a significant impact on some policy debates. The OTA, according to Plein and Webber, "played an active role in driving the course of debate in the area of biotechnology commercialization and international competition." In contrast, "the absence of support agency involvement in the controversial BST debate has perhaps contributed to policy consideration which has dragged on unresolved for half a decade." In the case of patents for developments in biotechnology, technology assessment adjusted incrementally as the new technology progressed. In general, congressional consideration of biotechnology is still in a formative stage. Congressional support agencies have done "a good job of identifying important aspects of biotechnology policy issues that warrant study."

Toward Improved Policy Analysis in Congress

Suggestions for how Congress might improve policy making for science and technology issues range from hortatory pronouncements about enlightened governing to the creation of new support agencies. In a recent report from a committee of the Carnegie Commission on Science, Technology, and Government (1991), for example, Congress was encouraged to establish a bicameral legislative service organization, patterned after the Environmental and Energy Study Conference. It would provide timely analysis of science and technology issues, facilitate dissemination of information, and devise a more direct way for members of Congress to receive "rapid-response technical analyses" from experts convened to assess specific legislative issues.

One option for Congress is to set up informal working groups around specific issues. These groups could be convened to give members and their staffs oral briefings and brief summary memos on major issues. The working group could debate policy options. They could review and critique drafts of proposed legislation. Members, of course, regularly consult with interest groups, but this ordinarily is done serially rather than in a collective setting, where members and staffers can learn from an exchange of view points among partisans. This kind of interaction often generates ideas that would not be produced in individual meetings. Staff and member time would be conserved. A more open deliberative process may be less candid, but it may be more useful in producing the best kind of policy proposals that are possible, given political limitations; it also could give members immediate feedback on the technical adequacy of their proposals and ideas about possible consequences. Statutes sometimes suffer from technical problems in drafting that could be identified by working group members (Carnegie 1991, 25–27).

This and other organizational innovations might be useful in addressing several of the problems identified here. Informal, ad hoc structural innovations can be fairly quickly organized and can help coordinate the interaction of overlapping committees without directly challenging established power structures or forcing members of Congress to give up existing committee assignments. Working groups can help facilitate the flow of information and help members sort out key arguments and underlying assumptions.

Second, members of Congress cannot escape difficult choices among competing values as they make policies for science and technology. But they can articulate more clearly these values, encourage public discussion of them, and help educate the public about the choices posed by scientific and technological advances. While there are few short-run political incentives in taking the time to explain complicated policy dilemmas to the public and engage in careful public debate, there could be great long-run benefits to Congress as an institution, and to society as a whole, if the public were to become more informed and sophisticated about the challenges and choices confronting policy makers, as democratic theorists going back to John Stuart Mill have suggested would occur. Members of Congress could use their access to the media to foster public information and literacy about emerging and future issues in science and technology. They could do much more to lead the debate in society about the risks of global climate change and other environmental problems, international competitiveness, the implications of human genome research, the challenges to privacy posed by new communications technologies, and the advantages and disadvantages of nuclear power. They could clarify budget choices and push debate beyond narrow, parochial concerns about the distribution of federal dollars.

This, of course, is an ambitious agenda for reorienting Congress in an era of weak parties and single-issue politics. But it certainly a direction that policy makers can move toward if they are serious about improving the policy making process.If members of Congress assumed a more active role in public discussion and debate, encouraged by congressional leaders through appropriate incentives, they would help produce the kind of policy leadership in their institution and elsewhere that has been lacking in some of the areas discussed in these chapters. They can engender a more sophisticated understanding of the limitations of scientific research and policy analysis, and the differences between political and technological choices. They can encourage a wide range of analyses and help make their results more visible and accessible to the public. In so doing, they can broaden the scope of policy analysis to address more thoroughly the range of political, social, and moral issues that are implicated in science and technology-related policies, as Vig illustrates, European parliamentarians are trying to do. They can help stimulate creative thinking that challenges accepted wisdom and encourages rethinking of accepted policy approaches.

A third need is to devise more effective ways of dealing with uncertainty. A variety of approaches and efforts are needed to expand the ability of policy makers to anticipate and respond to the challenges that lie ahead. One way of addressing this need is to look more carefully at the potential of good oversight in the legislative process. Congress can ensure that statutes include requirements, mechanisms, and resources for a regular, sustained assessment of how well statutes are working and when changes are needed. Assessment of how well statutes accomplish their goals and how relevant those goals are as problems evolve ought to be a major component of every law. While few members may relish revisiting in the near future contentious, difficult policy choices they have just made, effective government surely requires policy review and revision.

Oversight is often very contentious. Agency officials have chafed over intrusive oversight hearings. The more the executive branch recognizes and accept the role of Congress in overseeing its activities, the more likely that oversight relations will be productive. It is in the interests of members of both branches of government to improve the way in which Congress and the executive branch manage the separation of powers. Members of Congress need to be willing to invest the time required to develop expertise in the programs under their jurisdiction and to pursue with patience investigations and studies that sometimes take months and even years. They have to attract able professional staff and keep them long enough to provide continuity and perspective in oversight. And Congress must furnish the resources for thorough, systematic oversight activities that critically examines the adequacy of existing statutes.

Fourth, Congress can become a more future-oriented institution, but it must overcome the tremendous inertia that accompanies a collection of 535 independent individuals, working to protect their electoral fortunes within structures and procedures designed for the eighteenth and nineteenth centuries. Institutional improvements ought to be at the top of every congressional agenda as well as on the minds of political scientists and other observers of policy processes. Any such institutional changes could not be imposed from without, but would have to come from leadership committed to making Congress a more sophisticated, effective, and adaptive institution of governing.

These policy efforts are clearly among the most important undertakings of government, but how unique are they? How do they compare with the experience of other policy areas? Do we really have a clear understanding of how science and technology policies can be improved? If we do, what is the capacity of the policy making process for improvement? What is the relationship between formal analysis and more informal modes of learning, between rational assessment and self interest as motivators for change? The assessments in this book raise a host of these kinds of questions for future research. Given the importance of institutions like the Office of Technology Assessment, much useful work could be directed to studies of how such organizations function. Many other science and technology policy areas deserve exploration and study. The central role played by political variables in policy analysis points to the need for an in-depth comparison of science and technology policy in other countries can help improve our understanding of the relationship between political structures, public debate, and formal policy analyses.

Bibliography

Burns, James MacGregor. 1984. *The Power to Lead: The Crisis of the American Presidency* (New York: Simon and Schuster).

Carnegie Commission on Science, Technology, and Government. 1991. *Science, Technology, and Congress: Expert Advice and the Decision-making Process* (Washington DC: Carnegie Commission).

Chubb, John E. and Paul E. Peterson, eds. 1989. *Can the Government Govern?* (Washington DC: Brookings Institution).

Hawkesworth, M.E. 1988. *Theoretical Issues in Policy Analysis* (Albany, NY: SUNY Press).

Jasanoff, Sheila. 1990. *The Fifth Branch: Science Advisers as Policymakers* (Cambridge: Harvard University Press).

Marshall, Burke, ed. 1987. *A Workable Government? The Constitution After 200 Years* (New York: Norton).

Sundquist, James E. 1986. *Constitutional Reform and Effective Government* (Washington DC: Brookings Institution).

Sartori, Giovanni. 1962. *Democratic Theory* (Detroit, Mich.: Wayne State University Press).

Index